SpringerBriefs in Applied Sciences and Technology

PoliMI SpringerBriefs

Series Editors
Barbara Pernici, DEIB, Politecnico di Milano, Milano, Italy
Stefano Della Torre, DABC, Politecnico di Milano, Milano, Italy
Bianca M. Colosimo, DMEC, Politecnico di Milano, Milano, Italy
Tiziano Faravelli, DCHEM, Politecnico di Milano, Milano, Italy
Roberto Paolucci, DICA, Politecnico di Milano, Milano, Italy
Silvia Piardi, Design, Politecnico di Milano, Milano, Italy
Gabriele Pasqui, DASTU, Politecnico di Milano, Milano, Italy

Springer, in cooperation with Politecnico di Milano, publishes the PoliMI Springer-Briefs, concise summaries of cutting-edge research and practical applications across a wide spectrum of fields. Featuring compact volumes of 50 to 125 (150 as a maximum) pages, the series covers a range of contents from professional to academic in the following research areas carried out at Politecnico:

- Aerospace Engineering
- Bioengineering
- Electrical Engineering
- Energy and Nuclear Science and Technology
- Environmental and Infrastructure Engineering
- Industrial Chemistry and Chemical Engineering
- Information Technology
- Management, Economics and Industrial Engineering
- Materials Engineering
- Mathematical Models and Methods in Engineering
- Mechanical Engineering
- Structural Seismic and Geotechnical Engineering
- Built Environment and Construction Engineering
- Physics
- Design and Technologies
- Urban Planning, Design, and Policy

http://www.polimi.it

Giorgia Burzio · Venere Ferraro

Physicalizing Data: Design Approaches for Human-Data Interaction

Giorgia Burzio [ID]
Department of Design
Politecnico di Milano
Milan, Italy

Venere Ferraro [ID]
Department of Design
Politecnico di Milano
Milan, Italy

ISSN 2191-530X ISSN 2191-5318 (electronic)
SpringerBriefs in Applied Sciences and Technology
ISSN 2282-2577 ISSN 2282-2585 (electronic)
PoliMI SpringerBriefs
ISBN 978-3-032-05321-3 ISBN 978-3-032-05322-0 (eBook)
https://doi.org/10.1007/978-3-032-05322-0

© The Editor(s) (if applicable) and The Author(s), under exclusive license to Springer Nature Switzerland AG 2026

This work is subject to copyright. All rights are solely and exclusively licensed by the Publisher, whether the whole or part of the material is concerned, specifically the rights of translation, reprinting, reuse of illustrations, recitation, broadcasting, reproduction on microfilms or in any other physical way, and transmission or information storage and retrieval, electronic adaptation, computer software, or by similar or dissimilar methodology now known or hereafter developed.
The use of general descriptive names, registered names, trademarks, service marks, etc. in this publication does not imply, even in the absence of a specific statement, that such names are exempt from the relevant protective laws and regulations and therefore free for general use.
The publisher, the authors and the editors are safe to assume that the advice and information in this book are believed to be true and accurate at the date of publication. Neither the publisher nor the authors or the editors give a warranty, expressed or implied, with respect to the material contained herein or for any errors or omissions that may have been made. The publisher remains neutral with regard to jurisdictional claims in published maps and institutional affiliations.

This Springer imprint is published by the registered company Springer Nature Switzerland AG
The registered company address is: Gewerbestrasse 11, 6330 Cham, Switzerland

If disposing of this product, please recycle the paper.

Preface

Data is the new oil.

This is how mathematician Clive Humby defined data in contemporary society. While somewhat cynical, the phrase captures the economic and political weight that data holds today. It shapes decisions, influences behaviors, and drives global systems.

Despite this central role, data is often perceived as abstract, cold, and impersonal. Ultimately something disembodied and disconnected from everyday life. But data is never truly objective or "raw." Just like oil, it must be processed, refined, and contextualized to become valuable.

This book offers a critical reflection on the nature of data and its role in design, particularly within the field of interaction design, a discipline that connects digital and physical dimensions. Interaction designers frequently engage with data, capturing, processing, and communicating it through technology and materials. It is therefore essential to establish a dialogue between human-data interaction and the design practice.

The book explores how data is made tangible and sensorial, how it can be felt, touched, and experienced. Drawing on the principles of data physicalization, it presents insights from literature, highlights key case studies, and proposes new ways of engaging with data through material and aesthetic qualities. The case studies pay particular attention to how data is encoded or represented in artefacts and the types of user engagement they afford. Ultimately it paves out considerations to frame design interventions considering key principles in human-data interaction and emerging research in materials and interaction design.

Milan, Italy
Giorgia Burzio
Venere Ferraro

Acknowledgments We would like to warmly thank the students from M.Sc. Digital and Interaction Design, School of Design, Politecnico di Milano (AYs 2023–2024; 2024–2025) for their insightful contributions, creativity, and generous engagement in the research activities.

Competing Interests The authors have no competing interests to declare that are relevant to the content of this manuscript.

Contents

1 **Introduction** 1
2 **Data, Datum, Datafication** 3
   2.1   Data as Boundary Object 3
   2.2   Data and Self-tracking 5
   2.3   Data and Bodies 6
   2.4   Data Sense 7
   2.5   Conclusion 8
   References 9
3 **Grounding in Tangible Interaction Design** 11
   3.1   Framing Human-Data Interaction in Tangible Interaction 11
   3.2   Evolving Tangible Interaction: The Role of Shape-Changing Interfaces in Human-Data Engagement 12
   3.3   Embodied Interaction with Computing Things 14
   3.4   Related Areas: Soma Design and Interactive Materiality 16
   3.5   Conclusion 18
   References 19
4 **Data Physicalization in Theory** 21
   4.1   Situating Data Physicalization 21
   4.2   Literature Recognition 21
   4.3   Design Principles 23
   4.4   The Role of Metaphors When Representing Data 24
   4.5   Attention Points for Researchers 25
   4.6   Conclusion 26
   References 27

5	**Data Physicalization in Practice**		29
	5.1	An Annotated Portfolio of Data Physicalizations	29
	5.2	Vocabulary of Design Variables for Data Physicalization	35
	5.3	Feedback and Feedforward: An Analysis of the Annotated Portfolio	37
		5.3.1 Feedback	38
		5.3.2 Feedforward	39
	5.4	Applying the Variables Scheme in Speculative Artefacts	39
		5.4.1 Results	41
		5.4.2 Key Findings	41
	5.5	Conclusion	43
	References		44
6	**Looking Ahead: Which Emerging Directions?**		47
	6.1	A First Framework Attempt	47
		6.1.1 Variables Scheme	48
		6.1.2 Ideation Process	48
	6.2	Emerging Directions	50
		6.2.1 Living Data Mapping	50
		6.2.2 Lean Approach to Data	51
		6.2.3 Physicalizations and Data Ownership	51
	6.3	Final Remarks	52
	References		52

Chapter 1
Introduction

In a world increasingly shaped by data, designers are asking not just how to visualize information, but how to feel it. While data has become an important topic of investigation across many disciplines, it is only recently that began to attract deeper interest within the design community. Design is as a field deeply rooted in human experience and it is increasingly informed by more-than-human perspectives. Given this, it has a long tradition in mediating the relationship between humans and machines, and shaping technologies in ways that are accessible, usable, and engaging. This book sets out to explore how design can offer meaningful approaches for making data physically experienceable, with the goal of enhancing user experience and fostering an embodied sense of data.

The intention is to propose ways in which data can move beyond screens and be felt, touched, and interacted with in physical space. Through insights from literature and case studies this book considers how tangible data and physicalization processes can offer new perspectives and tools for designers. It aims to bring together the fields of human-data interaction and interaction design, proposing a shared vocabulary and approach for designing interactive systems and products that convey data through physical form.

The book is structured into five chapters. The first offers an overview of how data is understood in contemporary society, moving across different perspectives and challenging the view of data as abstract, objective, and neutral. Instead, it introduces data as situated, intimate, and entangled with human and technological contexts. The second chapter introduces the research landscape of tangible interaction design, including tangible user interfaces, shape-changing interfaces, and ICS (interactive, connected, smart) materials. The third presents the findings of an extensive literature review on data physicalization intersecting design and Human–Computer Interaction (HCI), outlining key principles and the role of metaphors in how we represent data. The fourth chapter focuses on data physicalization in practice, offering the results of an annotated portfolio activity. Here, examples of data physicalization and materialization are analyzed to draw out insights on interaction types, feedback, and feedforward.

Finally, the fifth chapter brings together the state of the art, organizing findings into a coherent model for future exploration through practice-based research. It also opens a space for reflection on the challenges and opportunities of physicalizing data, touching on emerging themes such as data mapping, living materials, and unpredictability.

Chapter 2
Data, Datum, Datafication

Abstract This chapter reframes data as a situated, embodied, and socio-technical construct rather than objective or raw. It draws on feminist and phenomenological perspectives to explore how data is entangled with bodies, technologies, and identities. Through examples like self-tracking and FemTech, it highlights data's personal and political dimensions.

2.1 Data as Boundary Object

The origin of the word data is from the Latin *datum*, meaning "something given". While originally "data" referred to givens or facts assumed to be true in logic, philosophy, and early science, contemporary research areas and domains are questioning the objectivity of data as *datum*, intending it instead as assemblages of things. In the past, the boundaries of data were confined to specific phenomena, deriving from extensive activities of observation, measurements, or experimentations. These days, the term big data tends to dominate how we talk about information, as it suggests scale and technological complexity. But data is not a single, fixed concept. Alongside big data, there is also small data, which is more accessible, often qualitative, and closely tied to human experience and interpretation. Data are mostly considered as representative, clearly reflecting the world and its measurement. Yet it can also be implied, where the absence of information speaks just as loudly as its presence. Sometimes data emerges through comparison, not from one dataset alone but from the relationship between many. Other times, meaning is shaped by metadata, the information that describes or gives context to other data. Data as a concept is a clear example of "boundary object". Formalized in science and technology studies (STS), a boundary object is an idea, tool, or artefact that is flexible enough to be interpreted differently by various communities, but robust enough to maintain a common identity across contexts (Star and Griesemer 1989).

In a nutshell, data has become increasingly pervasive into society's fabric, assuming different forms, not always clearly distinguishable. For instance, distinguish whether data are created by us or created about us has become quite difficult.

People generate themselves daily enormous quantity of data using their own devices, and therefore, the infrastructures and networks required to generate, store, and share data take the form of agents shaping technological and political worlds. According to Kitchin, who wrote an extensive book on the so-called "Data Revolution", what before was jealously guarded and traded at high costs, today is a *"wide, deep torrent of timely, varied, resolute and relational data that are relatively low in cost"* (Kitchin 2014).

Some says that data can help humans to take more informed, just decisions; nonetheless many highlight how data can also be means of exploitation, exclusion, and injustice. While there are certainly contrasting aspects, it is interesting to question the many ways data affect the way we think, act, and interact with the world. A whole research field called Human-Data Interaction (HDI) is precisely the emphasizing their human dimension of data. Meaning, how we shape them and how they shape us. Elmqvist introduced the term to describe how individuals interact with and make sense of large, often unstructured datasets. Building on this, Cafaro explores human–data interaction through the potential of context-aware and embodied forms of data engagement (Cafaro 2012). Other scholars have further expanded this perspective by examining the complex relationship between individuals and the pervasive collection and use of personal data in contemporary society (Mortier et al. 2014).

This process is defined in several ways, including datafication, so rendering aspects of everyday life into digitized information using digital technologies (van Dijck 2014). Taking a quite critical stance, we can say that commodifying personal data can lead to "dataveillance", both intentional and unintentional. In fact, people are mostly aware when they are granting their data to a certain company, as usually they obtain something in return. Nonetheless, researchers identify a "function creep", meaning ways in which datafication technologies have been used in situations far different from the ones originally intended and stated. Turning to the design field, Gaver and Boucher (2024) mapped how data underpin and are encoded in their own projects, distinguishing between captured and readymade data. Captured data is collected through sensors and technologies designed for specific purposes, while readymade data is repurposed from existing datasets.

A key matter, very close to us yet often ungraspable, are personal data. Those data are social interactions, financial behaviours, sexual preferences, health issues, biometric indicators, and so forth. Within the European context, a milestone event from 2018 is the European Union's General Data Protection Regulation (GDPR). The GDPR harmonized the existing privacy laws across the EU countries, regulating how data industries and big tech companies store and use personal data, to prevent misuse and unauthorized third-party access.

2.2 Data and Self-tracking

Personal data are often self-tracked, intentionally or unintentionally, for a variety of purposes. The rise of self-tracking as a socio-technical and cultural phenomenon began slowly in the early 2000s and accelerated rapidly after 2010 with the rise of Information and Communication Technologies (ICT), including mobile and cloud computing, social media, and Internet of Things (internetworked sensors and devices).

This growing trend of monitoring personal traits has far-reaching consequences. Deborah Lupton has extensively investigated the entanglements between bodies, digital technologies and data. She argues that data is neither objective nor raw; instead, it is often intimate and personal. Lupton uses the term "lively" to describe personal digital data. The term contains diverse interpretations. Digital data have their own life as they are used, reused and repurposed in the digital data economy. Moreover, they define aspects of human life, something alive, in evolution. It is obvious to say that the idea of raw, fixed data is not convincing. Ultimately, data are lively in the sense that they shape decisions and actions taken by people. So, even if data are perceived as abstract and objective matter by people, it is quite the contrary: data are malleable, intimate, imperfect and human by nature.

In her more recent book "Data Selves", Lupton presents the more-than-human entanglements involved in producing digital data, including how selfhood, software, hardware, and documentation work together to co-construct the "human-data assemblages" (Lupton 2020). Drawing from the vital materialism of Jane Bennett, data are active agents intertwined with the self, shaping the way we perceive ourselves and the way we are perceived by others. The instruments we use to measure phenomena (therefore produce data) do not exist independently the data itself. When we produce data, we perform what Karen Barad calls "agential cut": we arbitrarily separate the phenomena (measured objects) with the instruments we use for measuring. Even if agential cuts might seem like an ungraspable philosophical concept, it is quite recognizable and apparently unavoidable (Barad 1999).

While tech industries and startups dove into the self-tracking opportunity in the early 2000s, bottom-up movements and communities of tracker enthusiasts also began to spread, facilitated by the web. Quantified Self (QS) is an international, online movement of users and makers of self-tracking tools. Their website invites people *"interested to self-knowledge through numbers"* to share their insights and self-tracking habits with the community. An example of quantified data representations is given by Anna Franziska Michel, which uses her running and cycling data as material for her startling work in fashion design. The tracked data are processed by AI software and transformed into design patterns (Fig. 2.1).

This attempt showcases a clear "agential cut", where data are first simultaneously produced and recorded; then they are processed and rendered as numbers. As a follow up act for awareness, the owner of the data re-interprets the numbers into visual patterns for textile items. Such willingness and enthusiasm to track, monitor and represent personal data has been extensively examined both within Science and

Fig. 2.1 Anna Franziska Michel, using running and cycling data to inform my fashion

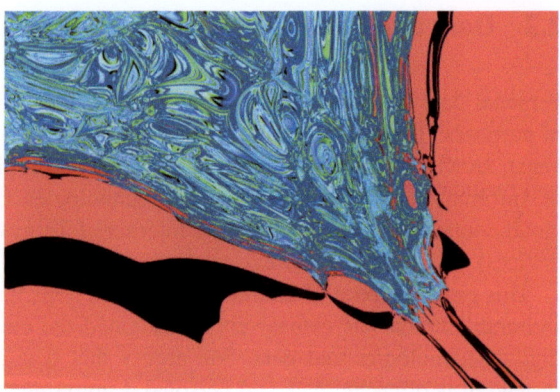

Technology Studies (STS) and Human–Computer Interaction (HCI) fields. People are willing to explore practices that can support the interpretation of data for a variety of purposes, including make them usable and knowable.

2.3 Data and Bodies

Among the plethora of personal data types, biodata is a critical subset. In fact, the relationship between bodies, data, and technology is complex and ambiguous. Zuboff (2015) refers to the datafication of bodies as "biovalue", highlighting how personal data enters society as a commercial commodity, like human tissue, blood, and cells. Thrift (2014) talks about hybrid beings, comprised of digital data and human flesh. Building on Donna Haraway's influential concept of the cyborg (Haraway 2010), Laura Forlano explores how data rituals experienced through a disabled body differs profoundly from that of able-bodied experience. Forlano traces the daily practices involved in managing Type 1 diabetes, showing how acts of measuring and quantifying are deeply intertwined with lived experience, care, and identity. In her account, data are not simply abstract or detached; they emerge as extensions of the body, situated within a technological and sci-fi landscape where the boundaries between human and machine, between bodies and data, become increasingly blurred (Forlano 2017).

Within the data-driven culture, female reproductive health has recently become a focus of attention for the MedTech industry. In 2016, Ida Tin, founder of the menstruation app Clue, coined the term "FemTech" (a blend of feminine and technology), referring to a wide range of products designed with women's digital health in mind. The industry is steadily growing, projected to be worth $50 billion by 2025. Several business magazines including Forbes reported that investors are now more willing than ever to support women in taking control of their health at every stage of life. FemTech applies to various areas such as women's health during the reproductive years, including ovulation sensing, pregnancy and breastfeeding, as well as during menopause. It also addresses conditions typically affecting female bodies, such as endometriosis and certain types of cancer. One common example of FemTech in practice is the growing use of digital and wearable devices for menstrual tracking. In 2019, Apple added a menstrual cycle tracking feature to its Health app, alongside other health indicators such as activity, heart rate, body measurements, sleep quality, and medication tracking.

A common critique is that such renewed interest into feminine health is mostly towed by venture capital opportunities and marketing. Filling the health gender gap undoubtedly holds positive aspects. In fact, such technologies are affording people to grasp traits and indicators otherwise imperceptible by observation and bodily sensors, alone. The opportunity to enhance such practices with technology might stimulate in-depth knowledge, awareness, and therefore, the motivation to act consciously in advance over certain problems. On the other hand, data are interpreted and leveraged to communicate a certain messaging, ambiguous and still rooted in a neo-liberal vision of "optimized bodies", fostering a sense of inadequacy for not being able to achieve certain standards promoted by tracking products. This aspect is also present in sport and fitness apps, which leverage aspects of gamification and persuasion to achieve certain results and self-optimize each one's performances. Perhaps the narrative around data and bodies should move away from the idea of data *to manage better* our health and well-being, focusing instead on data as tools for empowering us to make better decisions. While the difference seems fleeting, it holds radically different effects. Approaching health through the lens of optimization can foster feelings of inadequacy and anxiety, especially when individuals fall short of certain predefined standards. In contrast, many argue that using data to support self-awareness and emotional well-being, without the pressure to perform or conform, can promote a more empowering and compassionate understanding of health.

2.4 Data Sense

A still central aspect of the discourse on data representation (i.e., *to make data usable and knowable*, from above), is the fact that there is simply "too much data". Recent approaches in information design are prioritizing simplicity and minimalism in visualizations. In literature, the imperative of visualizing instead of overloading (Mengis and Eppler 2012) has been pointed out, yet it is still quite common to encounter messy dashboards overloaded with information. Referring again to the health and well-being domain, each trait, or biodata, is measured and reported through coloured graphs or numbers, often without being integrated with other key traits. This results in a fragmented overview of distinct data, difficult to read and to make sense of it.

A lean approach to data might be adopted when interacting with digital devices, such as the case of tracking technologies and digital sensors. Deborah Lupton reflects on how relating with personal data is a highly sensory experience, involving complex negotiations between the assessment of the tracked data and the appropriate response (Lupton 2013, 2014). She draws inspiration from the phenomenological research of Merleau-Ponty, which highlights the embodied dimension of "being in the world", with all our senses. Such phenomenological perspective backs up the conceptual shift from data literacy to data sense. Data literacy is often used to describe how people learn from data. It adopts a cognitive-centred approach, and it is widely used in domains such as informatics, education, information design. While data literacy is essential in certain contexts, the concept of data sense broads the scope, adopting an embodied perspective when learning from data. Data sense *"acknowledges the role of fleshy and affective bodily affordances in people's responses to data"* (Lupton 2020 pp 76). For these reasons, the data-interaction design space is expanding, challenging current conventions and proposing sensorial ways to engage with data, leveraging technologies and materials imbued with it.

Data sense allows a range of experiences activated by embodied processes of learning from data, including multiple senses, and therefore requiring methods that can meaningfully leverage bodily interactions with digital information.

2.5 Conclusion

This chapter has explored the evolving nature of data as a socio-technical construct and a boundary object. Data as matter of investigation moves across disciplinary fields such as STS, HCI, and design. Through historical, theoretical, and critical perspectives, the chapter has shown how data has shifted from its original notion as objective facts to more fluid, situated, and entangled assemblages involving people, technologies, environments, and infrastructures.

By drawing on feminist scholars such as Donna Haraway, Karen Barad, and Laura Forlano, along with the phenomenological perspective of Merleau-Ponty, the chapter highlights the embodied and relational dimensions of data. Data is not merely abstract or computational. It is lived, negotiated, and often deeply personal. From self-tracking practices to biodata and the development of FemTech, we have seen how data is both shaped by and shaping bodies, behaviours, identities, and social expectations.

The chapter also reflects on the concept of data sense by Deborah Lupton. This idea moves beyond the traditional focus on data literacy as a cognitive skill and invites a more embodied and affective way of engaging with data. It recognises that people experience data not only intellectually but also through sensory and emotional responses. This perspective encourages designers to rethink how data is communicated and understood.

Through these lenses, the chapter prepares the ground for what follows. The next chapters will explore how designers can create alternative forms of data representation that move beyond visual charts and graphs, opening space for material, spatial, and sensorial approaches that foster more inclusive and situated experiences of data.

References

Barad K (1999) Agential realism: feminist interventions in understanding scientific practices. The science studies reader, pp 1–11

Cafaro F (2012) Using embodied allegories to design gesture suites for human-data interaction. In: Proceedings of the 2012 ACM conference on ubiquitous computing, pp 560–563

Forlano L (2017) Data rituals in intimate infrastructures: crip time and the disabled cyborg body as an epistemic site of feminist science. Catal Fem Theory Technoscience 3(2):1–28

Gaver W, Boucher A (2024) Designing with data: an annotated portfolio. ACM Trans Comput-Hum Interact 31(6):1–25

Haraway D (2010) A cyborg manifesto (1985). Cultural theory: an anthology, p 454

Kitchin R (2014) The data revolution: big data, open data, data infrastructures and their consequences. Sage

Lupton D (2013) Quantifying the body: monitoring and measuring health in the age of mHealth technologies. Crit Public Health 23(4):393–403

Lupton D (2014) Self-tracking cultures: towards a sociology of personal informatics. In: Proceedings of the 26th Australian computer-human interaction conference on designing futures: The future of design, pp 77–86

Lupton D (2018) Lively data, social fitness and biovalue: the intersections of health and fitness self-tracking and social media. In: Thousand Oaks CA (ed) The sage handbook of social media. SAGE Publications, pp 562–578

Lupton D (2020) Data selves: more-than-human perspectives. Polity Press, Cambridge

Mengis J, Eppler MJ (2012) Visualizing instead of overloading: exploring the promise and problems of visual communication to reduce information overload. In: Information overload: an international challenge for professional engineers and technical communicators, pp 203–229

Mortier R, Haddadi H, Henderson T, McAuley D, Crowcroft J (2014) Human-data interaction: the human face of the data-driven society. arXiv:1412.6159

Star SL, Griesemer JR (1989) Institutional ecology,translations' and boundary objects: Amateurs and professionals in Berkeley's Museum of Vertebrate Zoology, 1907–39. Soc Stud Sci 19(3):387–420

Thrift N (2014) The 'sentient'city and what it may portend. Big Data Soc 1(1):2053951714532241

Van Dijck J (2014) Datafication, dataism and dataveillance: big data between scientific paradigm and ideology. Surveill Soc 12(2):197–208

Zuboff S (2015) Big other: surveillance capitalism and the prospects of an information civilization. J Inf Technol 30(1):75–89

Chapter 3
Grounding in Tangible Interaction Design

Abstract This chapter explores the evolution of human–computer interaction from screen-based paradigms to tangible and embodied forms of engagement with digital data. It discusses how Tangible User Interfaces (TUIs), shape-changing interfaces, embodied interaction, soma design, and interactive materiality offer more intuitive and sensory-rich ways of interacting with information.

3.1 Framing Human-Data Interaction in Tangible Interaction

The relationship between humans and digital information has undergone a profound transformation over the past few decades. Traditionally, digital interactions were confined to Graphical User Interfaces (GUIs), where information was displayed on screens and manipulated through indirect input devices such as mices and keyboards. However, research in Human–Computer Interaction (HCI) has paved the way for new paradigms that emphasize direct, physical engagement with digital data (Wiberg, 2014).

The introduction of Tangible User Interfaces (TUIs) provided a framework where digital information could be manipulated in a way that is analogous to physical objects. This shift set the stage for a broader understanding of embodied interaction, which considers how the human body and the physical world shape our computational experiences (Dourish 2001). More recently, fields such as soma design and interactive materiality have expanded these concepts, emphasizing movement, aesthetics, and materials that respond dynamically to interaction.

From a Human-Data Interaction (HDI) perspective, this evolution is crucial as it enhances the transparency, legibility, agency, and negotiability of data. HDI is concerned with how individuals understand, control, and engage with data that is collected about them. As physical and digital realms become more integrated, these interaction paradigms offer users more intuitive and meaningful ways to interact with data, ensuring that digital systems align with human cognition and behaviour. By exploring these paradigms, we gain a deeper appreciation of how digital and physical realms can be seamlessly integrated, ultimately enhancing the way we engage with information.

The relationship between humans and digital information has transformed significantly, shifting from screen-based interactions to more tangible and embodied experiences. Traditional digital interfaces relied on GUIs, whereas emerging paradigms such as TUIs, embodied interaction, soma design, and interactive materiality offer new ways of engaging with digital data.

3.2 Evolving Tangible Interaction: The Role of Shape-Changing Interfaces in Human-Data Engagement

Tangible user interfaces mark a fundamental shift in human–computer interaction, allowing users to manipulate digital data through physical objects. This paradigm was pioneered by Hiroshi Ishii and Brygg Ullmer in their seminal work Tangible Bits (1997), which proposed that digital information should not be confined to screens but should exist as graspable objects in the physical world. TUIs merge physical interaction with computational power, offering users a more intuitive and embodied way to engage with digital data.

A key principle of TUIs is the idea that computational data can be represented and controlled through tangible artifacts, enabling users to directly manipulate information through physical movements (Ishii 2008). This approach fosters deeper cognitive engagement by leveraging the human ability to grasp, move, and manipulate objects (Blackwell et al. 2007). Unlike traditional graphical user interfaces, which rely on abstract representations and indirect control mechanisms like mice and keyboards, TUIs create a direct mapping between physical actions and digital outcomes (Fishkin 2004).

One of the earliest examples of TUIs is metaDESK, a system that allowed users to interact with digital maps using physical objects. Another landmark project, Illuminating Clay, demonstrated how users could mold physical clay models while computational systems provided real-time feedback on structural and environmental simulations (Fig. 3.1). These projects highlighted the potential of TUIs to blend the physical and digital realms, offering an intuitive and engaging form of interaction.

TUIs have been applied in a variety of domains. In education, tangible programming tools such as Tangible Programming Blocks have been used to teach computational thinking through physical manipulation. Children engage with coding concepts by assembling physical blocks that represent different commands, making abstract programming logic more accessible and intuitive. This hands-on approach enhances problem-solving skills and promotes active learning.

In design and architecture, TUIs provide an innovative way to interact with 3D models. Urban planners, for instance, can shape physical models while computational tools analyze their designs in real-time, providing insights on environmental impact,

3.2 Evolving Tangible Interaction: The Role of Shape-Changing Interfaces ...

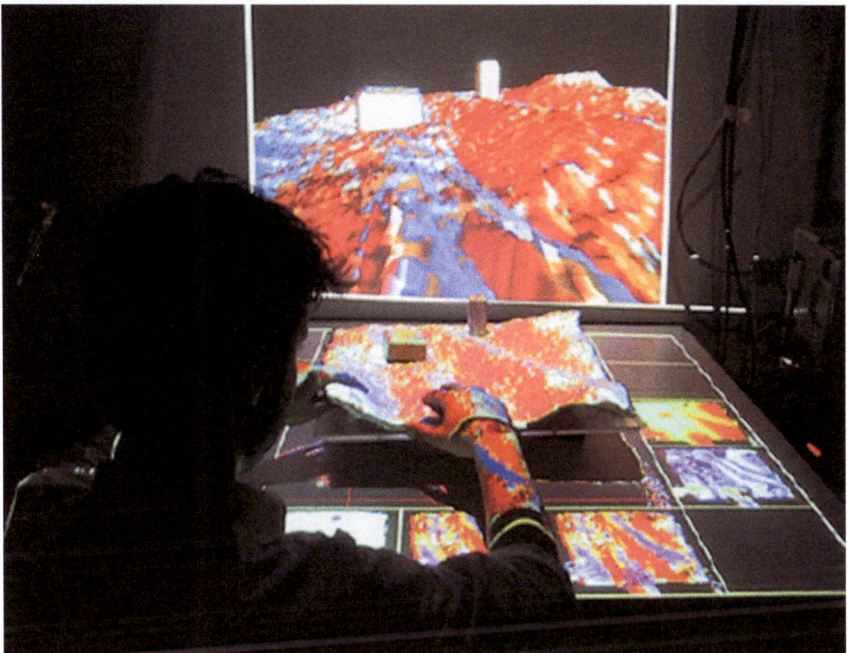

Fig. 3.1 Illuminating clay

spatial organization, and structural integrity. By integrating computational feedback with physical modelling, TUIs enhance creativity and allow for more informed decision-making.

From a human-data interaction (HDI) perspective, TUIs enhance legibility by allowing users to see and manipulate data in ways that align with their natural interactions with the physical world. Furthermore, agency is promoted as users gain direct control over how they interact with data through physical engagement. TUIs also support negotiability, allowing users to determine how data is manipulated and utilized across different applications.

Recent advancements in material science and computation have led to the emergence of shape-changing interfaces, an evolution of TUIs that further dissolves the boundaries between physical and digital interactions. Shape-changing materials introduce a new level of dynamic interaction where objects can morph in response to digital commands, altering their form, texture, or structure in real-time. These interfaces, as discussed by Coelho and Zigelbaum (2011), enable computational systems to extend beyond static input devices by integrating actuated surfaces capable of transformation. Such advancements are powered by smart materials such as shape-memory alloys (SMAs), dielectric elastomers, and magnetorheological fluids, all of which react to external stimuli to reconfigure their physical properties (Coelho and Zigelbaum 2011).

Rasmussen et al. (2012) provided a detailed taxonomy of shape-changing interfaces, categorizing them into different transformation types including orientation, form, volume, texture, and permeability changes. These transformations allow interfaces to dynamically adapt, providing a richer interaction space where users can not only manipulate objects but also receive tangible feedback from digital information. Unlike traditional TUIs, which rely on pre-defined tangible artifacts, shape-changing interfaces enable objects to serve multiple roles by reconfiguring their form on demand (Rasmussen et al. 2012).

The implications of shape-changing interfaces for HDI are significant. These interfaces enhance legibility by providing users with dynamic, multimodal feedback that better represents digital data. By allowing real-time reconfiguration, they support agency, enabling users to interact with and control computational systems in more nuanced ways. Negotiability is further extended as users can physically alter their interface to match their preferred mode of interaction, fostering a greater sense of control and personalization in digital experiences.

As research continues to explore new materials and actuation techniques, shape-changing interfaces will play a crucial role in the evolution of interactive computing, offering a seamless blend between the digital and physical worlds.

Ultimately, the integration of TUIs with shape-changing interfaces represents a major step toward more embodied, fluid, and responsive digital experiences. By embedding computation directly into materials, these interfaces bridge the gap between human cognition and machine interactivity, providing new opportunities for innovation across various domains, from education and design to accessibility and adaptive environments. As technology progresses, the challenge will be to refine these systems for scalability, usability, and broader real-world application while maintaining the principles of transparency, control, and meaningful human-data interaction.

3.3 Embodied Interaction with Computing Things

Embodied interaction builds on the foundations of tangible interfaces by emphasizing the interconnectedness of human cognition, physical movement, and the digital world. Paul Dourish introduced this concept in his book "Where the Action Is" (2001), arguing that computation should be understood as an embodied experience rather than as a purely abstract process. This perspective acknowledges that human understanding and interaction with technology are shaped by physical context, bodily movements, and environmental cues, which are crucial for human-data interaction (HDI).

A central principle of embodied interaction is cognitive offloading, where physical interaction reduces cognitive load by externalizing information processing. For example, a user manipulating a physical object to control a digital system does not need to mentally translate actions into abstract commands, as is often required in GUI-based interactions. Instead, the direct mapping between physical movements and digital responses simplifies the interaction process, making it more intuitive and effective. From an HDI perspective, this facilitates legibility, ensuring users can comprehend and engage with data in a more natural way, perhaps though the use of embodied metaphors.

Situated action is another key aspect of embodied interaction, which posits that human behaviour is context-dependent and shaped by environmental factors. Unlike traditional computational models that assume predefined steps and structured workflows, situated action recognizes that users adapt their behaviour dynamically based on real-world circumstances. This adaptability makes embodied interaction particularly relevant for complex problem-solving and collaborative tasks. In HDI, this means that negotiability is enhanced, as users can dynamically modify and control the way data is interpreted and used in different contexts.

Embodied interaction is widely applied in augmented reality (AR), virtual reality (VR), and interactive installations. In AR environments, digital information is overlaid onto the physical world, allowing users to interact with computational content through gestures, movements, and tangible objects. VR systems further immerse users by creating fully digital environments where physical actions directly influence virtual experiences. These technologies leverage embodied interaction principles to create immersive and engaging digital experiences while offering users more control over their data representation and engagement.

Tangible programming tools, such as AlgoBlock, (Fig. 3.2) exemplify how embodied interaction enhances learning and cognitive engagement. By allowing users to physically arrange and manipulate programming elements, these systems facilitate a more hands-on approach to coding.

Similarly, gesture-based interfaces, such as motion-tracking systems used in gaming and creative applications, provide intuitive and expressive ways for users

Fig. 3.2 A tangible programming language: AlgoBlock (Suzuki and Kato 1993)

to interact with digital content. In the context of HDI, these approaches contribute to agency, as they give users the ability to influence and control how data is manipulated through natural physical interactions. A recent example is Minimal Machines by Matters of Activity (2023), an augmented reality (AR) framework that capture data from the movements and actions of makers during construction. This data is then used to guide and adapt the building process in real time, creating a feedback loop between humans, materials, and machines.

As technology advances, embodied interaction is expected to evolve further with the integration of haptic feedback, biofeedback systems, and AI-driven adaptive interfaces. Haptic feedback enhances the sense of touch in digital interactions, making virtual objects feel more tangible. Biofeedback systems, which monitor physiological signals such as heart rate and muscle activity, could enable personalized and adaptive interactions based on users' physical and emotional states. AI-driven systems can analyze user movements and behaviours in real-time, adapting interfaces to provide more natural and intuitive interactions.

Ultimately, embodied interaction represents a paradigm shift in computing, moving away from traditional input–output models toward more fluid and intuitive engagement with digital systems. By recognizing the role of the body and environment in shaping human–computer interaction, embodied interaction opens new possibilities for more immersive, accessible, and meaningful digital experiences. From an HDI perspective, this shift enhances transparency, legibility, agency, and negotiability, ensuring that users have greater awareness and control over how their interactions generate, process, and utilize data in the digital world.

3.4 Related Areas: Soma Design and Interactive Materiality

Soma design and interactive materiality represent emerging frontiers in the field of interaction design, expanding on the concepts of tangible and embodied interaction by incorporating movement, aesthetics, and dynamic materials. These approaches emphasize the role of sensory experience and material responsiveness in shaping human–computer interaction and, by extension, human-data interaction (HDI).

Somaesthetic interaction design, or soma design, is rooted in the philosophical framework of somaesthetics, introduced by Richard Shusterman in 2000. This approach focuses on bodily awareness and the experiential dimensions of interaction.

Somaesthetics connects body and mind by rejecting the traditional dualist view that treats them as separate or opposed. Instead, it sees the body not just as a vessel, but an active participant in how we know, feel, and engage with the world. Kristina Höök and her collaborators have been instrumental in developing soma design methodologies, creating systems that encourage users to engage with technology in ways that heighten bodily awareness and sensory perception.

3.4 Related Areas: Soma Design and Interactive Materiality

From an HDI perspective, soma design contributes to agency by allowing users to understand and control the data generated through their interactions, ensuring a more personal and intuitive engagement with digital systems.

Interactive materiality challenges the assumption that materials are passive carriers of digital content. Instead, it explores how computation can be embedded into materials to create dynamic and responsive interactions.

Anna Vallgårda and Johan Redström introduced the concept of computational composites, which describe materials that integrate computational properties directly into their structure (Vallgårda and Redström 2007).

These materials enable new forms of interaction by changing their shape, texture, or properties in response to user input or environmental conditions (Vallgårda 2014). By making data legible and directly manipulable through material transformation, interactive materiality aligns closely with HDI principles.

Examples of interactive materials include shape-changing polymers that respond to electrical or thermal stimuli, computational textiles that alter their patterns based on digital signals, and self-assembling structures that morph in response to external forces (Fig. 3.3).

These innovations open new possibilities for interactive design, allowing for interfaces that are more fluid, adaptive, and integrated into the physical world (Jung and Stolterman 2011, 2012). Again, the integration of such materials with digital data enhances negotiability, as users can influence how data is represented and interacted with in a more natural and tangible manner (Fuchsberger et al. 2014).

As these fields continue to develop, they will redefine the relationship between humans and technology, creating more immersive and embodied digital experiences. By merging computation with bodily awareness and material responsiveness, soma design and interactive materiality push the boundaries of what is possible in interaction design, paving the way for a future where digital experiences are as tangible

Fig. 3.3 Example of computational concrete, courtesy of Glaister et al. (2007)

and dynamic as the physical world. In the context of HDI, these advancements will further enhance the way users perceive, understand, and engage with data, making digital interactions more seamless, intuitive, and human-centred.

3.5 Conclusion

The evolution of human–computer interaction has shifted significantly from the early days of screen-based interfaces to the current era of tangible and embodied interaction. The concept of Tangible Bits introduced by Ishii and Ullmer has had a lasting impact on the field, bridging the gap between the physical and digital worlds (Ishii and Ullmer 1997). Through tangible interfaces, we have seen how hands-on interaction fosters a deeper engagement with computational data, making complex digital systems more intuitive and accessible. The integration of physical and digital spaces has paved the way for innovative applications in education, design, and collaboration, proving that human interaction with data is most effective when it mirrors the way we naturally interact with our environment.

Embodied interaction has further expanded this paradigm, emphasizing the role of the human body in computational experiences. Unlike traditional interaction models that separate users from digital data, embodied systems consider physical movement, spatial awareness, and environmental context. This perspective not only improves usability but also creates richer, more immersive experiences. As seen in applications such as tangible programming and augmented reality, embodied interaction allows users to leverage their natural abilities, making digital interactions more meaningful and effective.

In parallel, shape-changing interfaces represent a natural extension of tangible interaction, introducing materials and objects that can dynamically reconfigure their form in response to digital inputs. By allowing physical interfaces to adapt and transform, they provide new ways of engaging with digital data, offering richer feedback and more flexible interaction possibilities. These interfaces align closely with the principles of human-data interaction (HDI), enhancing legibility, agency, and negotiability by enabling users to see and manipulate information through dynamic, responsive materials. When integrated with other interaction paradigms, shape-changing interfaces contribute to a more fluid and adaptive interaction space.

Soma design and interactive materiality represent the latest frontiers in this evolution. By focusing on the experiential and sensory aspects of interaction, soma design moves beyond usability to explore how technology can enhance bodily awareness and well-being. Interactive materiality, on the other hand, challenges the notion that materials are static, demonstrating how computational properties can be embedded into physical objects to create dynamic, responsive interactions. These approaches emphasize the importance of materiality, movement, and sensory feedback in shaping meaningful digital experiences, aligning with broader trends toward more embodied and immersive interactions.

As researchers and designers continue to explore these emerging fields, the boundaries between the physical and digital will become increasingly fluid, ultimately transforming the way we engage with information.

In conclusion, the evolution of tangible and embodied interaction underscores a fundamental shift in computing—one that moves away from traditional screen-based models toward more natural, intuitive, and human-centred interactions. The concepts discussed in this chapter highlight the importance of designing interfaces that align with the way humans think, move, and engage with the world. Whether through tangible user interfaces, embodied interaction, shape-changing materials, or somaesthetic design, the challenge ahead will be to integrate these innovations in human-data interaction smoothly, ensuring that digital experiences remain as intuitive, enriching, and responsive as our interactions with the physical world.

References

Blackwell AF, Fitzmaurice G, Holmquist LE, Ishii H, Ullmer B (2007) Tangible user interfaces in context and theory. In: CHI'07 extended abstracts on Human factors in computing systems, pp 2817–2820

Coelho M, Zigelbaum J (2011) Shape-changing interfaces. Pers Ubiquit Comput 15:161–173

Dourish P (2001) Where the action is. MIT press, Cambridge, p 28

Fishkin KP (2004) A taxonomy for and analysis of tangible interfaces. Pers Ubiquit Comput 8(5):347–358

Fuchsberger V, Murer M, Meneweger T, Tscheligi M (2014) Capturing the in-between of interactive artifacts and users: a materiality-centered approach. In: Proceedings of the 8th Nordic conference on human-computer interaction: fun, fast, foundational, pp 451–460

Glaister C, Mehin A, Rosen T (2007) Chronos chromos concrete. www.chromastone.com

Ishii H (2008) Tangible bits: beyond pixels. In: Proceedings of the 2nd international conference on Tangible and embedded interaction, pp 15–25

Jung H, Stolterman E (2011) Form and materiality in interaction design: a new approach to HCI. In: CHI'11 Extended abstracts on human factors in computing systems, pp 399–408

Jung H, Stolterman E (2012) Digital form and materiality: propositions for a new approach to interaction design research. In: Proceedings of the 7th Nordic conference on human-computer interaction: making sense through design, pp 645–654

Rasmussen MK, Pedersen EW, Petersen MG, Hornbæk K (2012) Shape-changing interfaces: a review of the design space and open research questions. In: Proceedings of the SIGCHI conference on human factors in computing systems, pp 735–744

Suzuki H, Kato H (1993) AlgoBlock: a tangible programming language, a tool for collaborative learning. In: Proceedings of 4th European logo conference, pp 297–303

Vallgårda A (2014) Giving form to computational things: developing a practice of interaction design. Pers Ubiquit Comput 18:577–592

Vallgårda A, Redström J (2007) Computational composites. In: Proceedings of the SIGCHI conference on human factors in computing systems, pp 513–522

Wiberg M (2014) Methodology for materiality: interaction design research through a material lens. Pers Ubiquit Comput 18:625–636

Ishii H, Ullmer B (1997, March). Tangible bits: towards seamless interfaces between people, bits and atoms. In Proceedings of the ACM SIGCHI Conference on Human factors in computing systems, pp. 234–241

Chapter 4
Data Physicalization in Theory

Abstract This chapter explores data physicalization as an interdisciplinary design practice that transforms abstract data into material, embodied, and sensory experiences. It outlines the theoretical foundations of the field, emphasizing principles such as embodiment, temporality, materiality, and metaphors.

4.1 Situating Data Physicalization

As we increasingly live in a datafied world, where numbers and metrics underpin decisions in domains as varied as climate action, public health, education, and personal well-being, the question of how we perceive, interpret, and interact with data becomes central to both technology and design. While data is commonly visualized through graphs, charts, and dashboards, these representations often remain abstract, detached, and cognitively taxing, especially for non-expert audiences. Data physicalization offers a radical shift in this paradigm: it proposes the materialization of data through physical forms, engaging users in embodied, multisensory, and often participatory ways. Emerging at the intersection of human–computer interaction, design, and information visualization, data phys aims not only to make data more accessible but to transform it into something that can be lived with, reflected upon, and emotionally connected to. This chapter delves into the theoretical foundations of physicalization, exploring its origins and evolution, core design principles, the crucial role of metaphor in physical data representation, and key challenges for researchers and practitioners.

4.2 Literature Recognition

Recent years have seen an increase in scholarly efforts to define and map the boundaries of data physicalization as a research field. As a domain that intersects multiple disciplines, including interaction design, HCI, digital fabrication, and information visualization, the literature reveals a rich variety of epistemological perspectives. Traditionally, data representation in HCI followed visual paradigms emphasizing clarity, accuracy, and legibility. However, the physicalization movement reorients this approach by incorporating bodily engagement, materiality, and situated interaction into the processes of data encoding and decoding.

One of the most influential definitions was articulated by Jansen et al. (2015), who define data physicalization as *"a physical artefact whose geometry or material properties encode data"*. This foundational contribution sparked a wide array of related concepts and taxonomies, including embedded data representations (Willett et al. 2017), autographic visualization (Offenhuber 2020), data manifestations (von Ompteda 2019), data agents (Karyda et al. 2021), and data materialization (Starrett et al. 2018). Each of these terms emphasizes a different aspect of the data-object relationship, whether focusing on processual emergence, material agency, or interpretive openness.

Rather than presenting data as objective, universal, and easily decodable, many researchers emphasize the situatedness and contextual contingency of data representation. Drawing from information studies and science and technology studies (STS), authors such as Offenhuber argue that data is not neutral or detached but materially inscribed and entangled with the processes that generate it. His concept of autographic visualization highlights how environmental traces such as rust patterns, wear marks, or physical deformations can be read as data, shifting the emphasis from symbolic to indexical representation.

Sanches et al. (2022) further enrich this epistemological turn by drawing on Karen Barad's theory of agential realism. They propose that data does not pre-exist its representation but co-emerges through interactions between humans, materials, and technologies. This entangled view is particularly important for understanding physicalizations as not merely representational but performative: artefacts that generate new understandings and experiences through their material and spatial arrangements.

Systematic efforts to map the field have been carried out by Burzio and Ferraro (2024), who clarified physicalization areas according to user interaction and data encoding level. Their work identifies a growing body of contributions organized across theoretical, methodological, and technological dimensions. They also underscore a shift toward embodied and participatory approaches, where physicalizations are situated within lived contexts and social practices.

Frameworks such as physecology (Sauvé et al. 2022) represent attempts to formalize the parameters of this emerging field. Physecology examines how physicalizations operate within specific environments, audiences, and temporalities. It moves beyond static representations to examine the lifecycle, interaction dynamics, and spatial embedding of data artefacts. These insights are critical for understanding the expanded design space in which data phys operates.

In summary, the literature on data physicalization reflects a broader shift in how data is conceptualized, designed, and experienced. From foundational definitions to more speculative and critical approaches, the field is increasingly characterized by its interdisciplinarity, its commitment to embodied and situated engagement, and its recognition of data as a cultural, material, and social phenomenon. This transformation lays the groundwork for novel design practices and research methodologies that challenge conventional paradigms of data interaction.

4.3 Design Principles

Building upon the conceptual foundations laid out in the literature, the next crucial step is to examine how these ideas translate into practical and methodological frameworks. The design of data physicalizations involves more than simply choosing a form or material, requiring careful attention to the ways in which data is encoded into physical media, how interaction is structured, and how context and participation shape user experience. In the following section, we explore the principles that underlie these decisions, drawing attention to how material, sensory, and social dynamics influence the creation and interpretation of physicalized data.

Designing data physicalizations involves translating abstract information into material forms that not only convey meaning but engage users in perceptual, embodied, and often participatory experiences. As the field evolves, researchers and practitioners have started to identify key principles that guide this translation process, moving beyond the visual conventions of data visualization to a broader, more inclusive, and sensorially rich design space.

One of the central principles emerging from the literature is the importance of embodiment in interaction. Rooted in phenomenological theories and developed within HCI, embodied interaction shifts the focus from disembodied cognition to engaged, lived experience. Data physicalizations that leverage embodied schemas—such as up/down, in/out, or near/far—draw on bodily patterns that structure how people intuitively understand and navigate data. These patterns are not only cognitive but sensorimotor, and their integration into physicalization design allows users to make sense of complex information through tactile, spatial, and kinesthetic modalities.

Another principle concerns the diverse material properties and behaviours that can be harnessed in the design of physicalizations. Materials are not passive substrates but active agents in the communication of data. The choice of material, being it wood, textile, biopolymer, or smart material, influences not only functionalities and affordances of the artefact but also its symbolic and emotional resonance. Emerging materials, such as responsive polymers, soft robotics, and humidity-reactive films, offer new possibilities for creating data-driven artefacts that are not static but dynamic, adaptive, and even alive.

Temporality is also increasingly recognized as a key design variable. Physicalizations unfold across time: some change in real-time in response to data streams; others evolve through user interactions, while others capture traces of past events. Vallgårda's concept of temporal form provides a useful framework for understanding these dynamics, emphasizing how time shapes not only the behaviour of artefacts but the meaning users derive from them (Vallgårda et al. 2015). Designers must thus consider not just what a physicalization looks like, but how it changes, ages, decays, or grows.

Finally, the literature emphasizes the value of participatory and co-design approaches in the creation of physicalizations. Rather than being designed solely by experts, many physicalizations now emerge from collaborative processes involving users, communities, or stakeholders. These co-created artefacts reflect lived experiences, collective values, and shared concerns, making data not just more accessible, but more meaningful. Such approaches are particularly relevant in civic, environmental, and health contexts, where the aim is not merely to inform but to empower people and communities.

Together, these principles—embodiment, material agency, temporality, and participation—form the backbone of contemporary design practice in data physicalization. They point toward a paradigm where data is not only seen or understood, but felt, inhabited, and acted upon.

Building on these foundational design principles, it becomes essential to consider the conceptual strategies that facilitate the translation of abstract data into forms that resonate with human perception and understanding. Rather than translation, the notion of transduction may offer a more nuanced way to understand how data takes shape across sensory modalities. Gilbert Simondon, a French philosopher, used the term transduction in the 1950s in a metaphysical and epistemological sense to describe how structures and knowledge emerge and evolve. In cognitive science and design, it refers to the process of converting abstract data into sensory signals, including light patterns, soundscapes, tactile vibrations, and more (Ammar-Khodja and Celerier 2025).

One of the most powerful of these strategies lies in the use of analogical thinking—ways of structuring meaning through familiar frameworks drawn from embodied and cultural experience. Among these, metaphor offers designers a rich toolkit for crafting physicalizations that are not only legible but affectively and cognitively engaging. These metaphorical logics help bridge the gap between abstract numerical or symbolic data and the lived, sensory experiences of individuals. They shape how physicalizations are interpreted and interacted with, offering cognitive shortcuts, aesthetic cues, and cultural references.

4.4 The Role of Metaphors When Representing Data

Metaphors can be considered indispensable tools in data physicalization, serving as conceptual bridges between abstract data and concrete physical experience. Unlike visualizations that often rely on standardized encodings (e.g., bar heights or pie slices), physicalizations draw from a much broader semiotic repertoire. Here, metaphors are not simply illustrative but determinant as they shape how users perceive, interact with, and derive meaning from data.

The literature identifies several kinds of metaphoric strategies. Some physicalizations adopt metaphors that are close to both data and reality, employing familiar forms with direct mappings. For example, a wearable that inflates with increased heart rate metaphorically enacts the sensation of bodily exertion. Others adopt forms that are metaphorically close to reality but distant from the data, such as trees or clouds, which serve as evocative but ambiguous representations. Still others use abstract geometries with minimal metaphorical cues, prompting users to actively construct meaning through interaction.

Research by Zhao and Vande Moere categorizes these metaphors along axes of interpretability and familiarity, helping designers to calibrate the balance between abstraction and intuitiveness (Zhao and Vande Moere 2008). Meanwhile, Bakker and colleagues emphasize the use of embodied metaphors grounded in physical experience that structure human understanding from infancy. These schemas include containment, flow, and balance, and they can be leveraged to create intuitive interactions with data even in the absence of explicit instruction (Bakker et al. 2011).

Metaphors also play a critical role in framing the emotional and political resonance of physicalizations. A slowly decaying artefact that reflects air pollution levels can evoke a sense of loss or urgency, making the data not only visible but affectively compelling. Conversely, a modular installation that invites community members to contribute physical tokens to represent their energy use may foster a sense of collective responsibility.

Importantly, metaphors in data physicalization are not static but dynamic. As users interact with the artefact, their understanding of the metaphor (and by extension, the data) can shift. This interpretive openness is a distinctive strength of physicalization, allowing for personal, contextual, and evolving engagements with data.

4.5 Attention Points for Researchers

As the field of data physicalization expands, it brings with it several critical challenges and research questions. These attention points highlight the areas where further inquiry, methodological innovation, and critical reflection are most needed.

One key issue concerns the vocabulary of variables used in data physicalization. Traditional visualization frameworks often rely on encoding variables such as position, size, colour, etc., that assume stable, symbolic mappings between data and representation. However, physicalizations increasingly incorporate what Hornecker et al. describe as implicit properties and consequential aspects, namely the unintended, yet determinant, effects of the data variables adopted. This includes material symbolism, spatial placement, interaction style, and even the side effects like shadow or weight. Researchers have begun to propose hybrid frameworks, such as the variables scheme, to better account for these multimodal, ephemeral, and embodied forms of meaning-making.

Time is another area requiring deeper exploration. How long is a data physicalization meaningful? When does a dynamic form become a static artefact? And how do users interpret changes that unfold over seconds, minutes, or months? Concepts like feedback and feedforward, while useful per se, may be insufficient to describe the temporal complexity of physicalization. This central aspect is further approached and elaborated in the following chapter.

The question of meaningfulness itself presents a conceptual challenge. Scholars such as Mekler and Hornbæk argue for a richer evaluative framework that includes dimensions like coherence, connectedness, significance, and resonance (Mekler and Hornbæk 2019). These criteria move beyond usability to consider how data artefacts support reflection, identity formation, and emotional engagement. Karyda and colleagues build on this by exploring how physicalized representations of personal data can act as mirrors, prompts, or provocations, fostering new understandings of the self and world.

Finally, the integration of emerging materials—bio-based, responsive, or even living—opens both opportunities and uncertainties. How do we design for materials that age, mutate, or behave unpredictably? What are the ethical implications of using living systems to encode data? How do we evaluate the interpretability of artefacts whose behaviour is co-determined by environmental conditions? These questions point to a frontier where data design intersects with material science, synthetic biology, and critical design. In short, the challenges of data physicalization are not just technical but conceptual, ethical, and experiential. They invite us to rethink what data is, how it should be represented, and what it means to live with data in physical, shared, and responsive ways.

4.6 Conclusion

Data physicalization is a rapidly maturing field that redefines the relationship between people and data through the lens of materiality, embodiment, and experience. Rather than reducing data to abstract visualizations or numerical dashboards, data phys enables new ways of living with, making sense of, and acting upon data. It expands the possibilities of interaction beyond screen and symbol, embracing bodily, temporal, affective, and social dimensions (Lupton 2018). This chapter has shown that data physicalization is deeply interdisciplinary, drawing insights from HCI, design research, STS, and material studies. Its theoretical foundations are grounded in critical re-evaluations of what constitutes data, who has agency in its production and representation, and how it is embedded in the material world. In doing so, physicalization challenges dominant paradigms of clarity, neutrality, and legibility, instead opening space for ambiguity, reflexivity, and co-creation.

Design principles in data phys emphasize the value of embodied interaction, material expressiveness, participatory processes, and temporal sensitivity. Metaphors play a crucial role in making abstract data relatable and emotionally resonant, while emerging materials invite speculative and situated engagements with data as something dynamic and evolving.

Key challenges remain: the development of shared vocabularies and evaluation frameworks, the need for inclusive design practices, and the ethical implications of working with sensitive, performative, or responsive data types. Yet these are not barriers, but invitations. They call for new alliances between disciplines, new methods of inquiry, and new imaginaries of what it means to design with data.

References

Ammar-Khodja B, Celerier JM (2025) Sensing beyond graphs, artifacts, and matter: data polysensualization as a framework for environmental detection in art and design practices

Bakker S, Antle AN, Van Den Hoven E (2011) Embodied metaphors in tangible interaction design. Pers Ubiquit Comput 16(4):433–449

Burzio G, Ferraro V (2024) Clarifying definitions: a scoping review of data physicalization in human-computer interaction. In: Proceedings of the 18th multi conference on computer science and information systems (MCCSIS)

Hornecker E, Hogan T, Hinrichs U, Van Koningsbruggen R (2023) A design vocabulary for data physicalization. ACM Trans Comput-Hum Interact 31(1):1–62

Jansen Y, Dragicevic P, Isenberg P et al (2015) Opportunities and challenges for data physicalization. In: Proceedings of the 33rd annual ACM conference on human factors in computing systems (CHI '15), pp 3227–3236

Karyda M, Mekler ED, Lucero A (2021) Data agents: promoting reflection through meaningful representations of personal data in everyday life. In: Proceedings of the 2021 CHI conference on human factors in computing systems (CHI '21), Article 168

Lupton D (2018) How do data come to matter? Living and becoming with personal data. Big Data Soc 5(2):1–11

Mekler ED, Hornbæk K (2019) A framework for the experience of meaning in human-computer interaction. In: Proceedings of the 2019 CHI conference on human factors in computing systems (CHI '19), Article 243

Offenhuber D (2020) What we talk about when we talk about data physicality. IEEE Comput Graphics Appl 40(6):25–37

von Ompteda K (2019) Data manifestation: merging the human world and global climate change. In: IEEE VIS arts program (VISAP)

Sanches P, Howell N, Tsaknaki V, Jenkins T, Helms K (2022) Diffraction-in-action: designerly explorations of agential realism through lived data. In: Proceedings of the 2022 CHI conference on human factors in computing systems (CHI '22), Article 216

Sauvé K, Sturdee M, Houben S (2022) Physecology: a conceptual framework to describe data physicalizations in their real-world context. ACM Trans Comput-Hum Interact 29(3):1–33

Starrett C, Reiser S, Pacio T (2018) Data materialization: a hybrid process of crafting a teapot. In: ACM SIGGRAPH 2018 art gallery, pp 381–385

Vallgårda A, Winther MT, Mørch N, Vizer EE (2015) Temporal form in interaction design. Int J des 9(3):1–15

Willett W, Jansen Y, Dragicevic P (2017) Embedded data representations. IEEE Trans vis Comput Graph 23(1):461–470

Zhao J, Vande Moere A (2008) Embodiment in data sculpture: a model of the physical visualization of information. In: Proceedings of the 3rd international conference on digital interactive media in entertainment and arts (DIMEA), pp 343–350

Chapter 5
Data Physicalization in Practice

Abstract This chapter presents an annotated portfolio of data physicalizations, examining their forms, functions, and user interactions across diverse domains. Drawing on established frameworks and new taxonomies, it analyzes artefacts through dimensions such as encoding variables, interaction anatomies (static, constructive, dynamic), and the feedback/feedforward model.

5.1 An Annotated Portfolio of Data Physicalizations

Data physicalization is a burgeoning area within design and data representation, blending tangible artefacts with encoded information to create meaningful interactions (Hogan and Hornecker 2013). The physical, actual manifestation of a physicalization in the real world sits closely to diverse fields. Giving its explicit intention to allow nuanced and sensorial interaction with data, one of these fields is surely Human–Computer Interaction (HCI). Another is fabrication and product design, considering that data phys can take the shape of everyday objects equipped with novel meanings, referring to datasets (Sosa et al. 2018). Ultimately it strongly links to information visualization, as the datasets might be gathered, curated, and re-arranged to be communicated in an effective way to convey the right message or story.

The theoretical foundations have provided a robust starting point to understand and define data phys as a research field, including its scope and challenges, and in relation to HCI, fabrication and information visualization. Nonetheless, a scouting into real-world examples became essential to fully grasp the diverse applications and possibilities of physicalizations. To address this gap, this chapter presents a collection of practical cases to give insights on how a specific type of artefact that physicalizes data might look like. Defined as annotated portfolio this method serves as a comprehensive collection of practical cases that illustrate the breadth and depth of the data phys potential. Annotated portfolios derive conceptualizations and insights from practical cases and products, by abstracting the actual features of the device

(Gaver and Bowers 2012). This portfolio not only highlights the physical embodiments of data but also underscores their varying purposes, techniques, and contexts of use.

As extensively discussed, data physicalization is a research area encompassing diverse sub-areas. Besides visualization, there is data edibilization, defined as the encoding of data through edible materials (Wang et al. 2016), and data sonification, defined as the encoding of data through sound (Kaper et al. 1999), to mention a few. Moreover, another sub-area is data art, which focuses on representing links between data and the artistic creation. Data art is closely link to the foundational concept of data sculpture, data-based artefacts possessing both artistic and aesthetic qualities (Zhao and Vande Moere 2008).

After careful reflection, we excluded from the portfolio tracking devices, which are monitoring and displaying data. We decided to exclude cases of data edibilization and data sonification as well. These artefacts did not directly respond to Jansen's definition (2015) of *"physical object whose geometry or material properties encode data"*, which informed us in the cases collection. Examples of physicalizations included are sound datasets encoded in ceramic cups (Desjardins and Tihanyi 2019); slow feedback mechanism to monitor sport performances (Menheere et al. 2021); data strings constructed by the public (Domestic Data Streamers 2013); and a dynamic menstrual cycle calendar (Shirley Wu 2023).

The process of annotating case studies was an iterative one. At first, the case studies were gathered in a Figma board and summarized into visual cards. Each card included the following annotations: (i) artefact's title and author(s); (ii) pictures; (iii) typology of artefact; (iv) technique used; (v) aim.

Concerning the typology of artefacts, we used macro-categories such as product, product system (in case of a physical product and a digital one, or a service), installation (in case of a museum exhibit, for instance), and material technology. As for the technique used, we assumed a similar approach of macro-categories, since a deep review of the methods and tools was not a topic of interest at first. We instead aimed at understanding through which techniques data are being physicalized into an artefact, might it be a product, installation and so on. The annotations were digital fabrication, sensors and actuators, artificial intelligence, motion tracking, shape-changing materials, bioactive materials.

As for the aim, we mainly used the ones provided by Jansen et al. (2015) and Dragicevic et al. (2020): raise awareness, communication of complex information, increase understanding and cognition, create an artistic piece, and support decision-making. Two case studies could not be annotated by any of those; thus, we added gathering data and persuade behaviours.

While collecting and annotating cases, we acquired insights on the different anatomies of data phys, annotating them into static, constructive, and dynamic. These categories were extracted from theoretical literature and further conceptualized, and validated, through the case studies. Specifically, a static data phys do not update or change in response to new data inputs. It can be defined, in other words, as a physical representation of a certain dataset in a certain time, useful for contemplation, comprehension, awareness, or merely aesthetic purposes. Constructive data phys emphasizes the construction, manipulation, and assembly of physical artefacts that represent data (Wijers et al. 2024). Such physicalizations can be re-arranged from time to time according to the dataset, and they strongly rely to metaphors while representing data. This since each coin, or token, represents a certain data type that has been determined beforehand. Moreover, indicators such as token dimension, amount, colour and shape might indicate data connotations, acting as variables. Ultimately, dynamic data phys update based on a new datasets' input (Signer et al. 2018). This typology relies on techniques and materials which can change shape according to certain inputs, both naturally (in case of chemical compounds, bioactive or smart materials), digitally, using electronic prototyping, or a combination of both. Data enact certain material behaviours so that they represent the underlying data in an embodied, aesthetic, and interactive way.

According to its anatomy, the interaction between user and data physicalization changes consistently, as it allows for different purposes as well as a level of freedom. For instance, constructive physicalizations afford a collective engagement, while static ones afford deeply an individual one. Constructive data phys can be more suited to support decision-making, while dynamic ones might be ideal for communicating data that change daily, such as environmental data.

We annotated whether the physicalization is intended for personal or collective use. The following Table reports six case studies as examples of the annotation's methodology (Table 5.1).

From the annotated portfolio it emerges that most cases are static physicalizations, using digital fabrication as a technique. This is perhaps the most effective and established way to represent data into tangible form: take a dataset, filter and transform it to a code language such as g.code (computer numerical control and 3D printing programming language), and select the most appropriate technique to convert this language into physical shapes. These techniques are often 3D printing or CNC machinery. A process is described by Desjardins and Tihanyi (2019) in crafting ListeningCups. Another detailed process is described by Pennerup Nilsson (2023), which recycled unused datasets into novel shapes, reasoning on aesthetics and the formal value of data. Those datasets came from a variety of different available sources, such as root canals, library trips, and recreational fishing. Nilsson's work highlights an emerging aspect of treating data as a material for design: the reuse and sustainability of datasets. It draws attention to the vast amount of data generated and then discarded without being used. By applying design methodologies that repurpose such datasets, it becomes possible to assign new meaning to seemingly useless data and to explore novel aesthetic and expressive possibilities.

On the other hand, constructive and dynamic data physicalizations do not yet appear to be fully formalized—either aesthetically or in terms of interaction—and are perhaps still regarded as stylistic exercises.

Table 5.1 Case studies annotation, six examples

Image	Name and description	Domain	Technique	Aim	Anatomy
	ListeningCups Audrey Desjardins Timea Tihanyi A set of 3D printed porce-lain cups em-bedded with datasets of eve-ryday ambient sounds.	Product	Digital fabrication	Increase understanding and cognition	Static, personal phys
	The City Lab: Letting the city talk about perception (Domestic Data Streamers 2016) The City Lab asked the audience a series of questions using different laboratory devices that collected the answers in liquid form.	Installation	Re-use of objects from another purpose	Gathering data, create an artistic piece	Constructive, collective phys

(continued)

5.1 An Annotated Portfolio of Data Physicalizations

Table 5.1 (continued)

Image	Name and description	Domain	Technique	Aim	Anatomy
	SENSBIOM (crafting plastics! & DumoLab Research, 2023) Presented at Milan Design Week in 2023 SENSBIOM is a collection of biopolymer lattices able to change colour evidencing UVR making visitors aware of invisible threats.	Installation	Bioactive materials, digital fabrication	Raise awareness, create an artistic piece	Dynamic, collective phys
	Air Transformed Better with Data (Quick & Posavec 2015) Wearable data objects that communicate air pollution based on open air quality data from Sheffield, UK.	Product	Digital fabrication	Raise awareness, communication of complex information	Static, personal phys

(continued)

Table 5.1 (continued)

Image	Name and description	Domain	Technique	Aim	Anatomy
	Laina (Daphne Menheere et al. 2021) Shape-changing furniture presenting physicalized running data through a slow feedback mechanism.	Product	Digital fabrication, sensors	Support decision making, persuade behaviour	Dynamic, personal phys
	Motiis (Pepping et al. 2020) A system for parents that measures and tangibly visualizes children's emotions experienced during gaming sessions.	Product	Digital fabrication, sensors, haptics	Raise awareness, support decision-making	Dynamic, personal phys (children as secondary users)

5.2 Vocabulary of Design Variables for Data Physicalization

When working with data as design material, we might look at design variables to shape the feedback and feedforward—therefore the interaction—of the physicalization. In any data communication activity, encoding variables (i.e., the properties of the material used to encode data) are a key design dimension, even more for immersive and multisensory data representations. In other words, encoding variables are the physical attributes (e.g., shape, texture, size) that can be manipulated to represent data in physical forms (Ranasinghe and Degbelo 2023).

Paneels and Roberts (2009) provided a review of design variables for haptic data visualization, exploring how haptic, touch-based feedback could be leveraged to represent data. They categorize the designs into distinct classes based on the type of data, the mode of interaction, and the purpose of the visualization. Even if the authors refer to haptic data representations, their framework proves very useful and suited to be adapted and integrated for data physicalization designs. They identified data types, interaction modes, and visualization purposes (Table 5.2).

The sensory feedback categories from the Panëels and Roberts model were not fully adequate for data physicalization. In fact, from the annotated portfolio several more variables emerged, in addition to force feedback, vibration feedback and texture simulation, which are accurate for haptic data representations. Ranasinghe and Degbelo (2023) provided an exhaustive review of encoding variables for data physicalization, classified in physical, visual, haptic, sonic, olfactory, gustatory, and dynamic ones. Such review, besides offering a comprehensive classification of variables, proved itself useful to pair and map our own case studies from the annotated portfolio according to the variable used to encode data (Fig. 5.2). We included colour value, hue and saturation; vibration force; tangible location, elevation, size, texture, arrangement, shape, numerousness, orientation.

Ranasinghe and Degbelo also introduced the concept of dynamic indicators to highlight how variables change over time. One such indicator is temporal order, which describes the sequence of state transitions from one value (e.g., A) to another (e.g., B). This change can itself be a part of the data physicalization process.

For instance, let's consider the Sensbiom installation by design studio craftingplastics!, which uses bioactive materials to change color in response to UV levels in real-time. The variable in this case is the colour, with states that range from yellow to orange to red, including various intermediate shades. The temporal order represents the transition between these states, such as from yellow to orange. The duration of this transition communicates the speed of the UV level changes: a shorter transition indicates rapid changes, while a longer transition indicates slower changes.

The scheme (Fig. 5.1) encompasses key variables for physicalizing data and three key dimensions to consider shaping the physicalization process. First the kind of data, which could potentially guide the variables choice. Then the dynamic indicators, described above, which frame the interaction from a temporal point of view. Lastly the visualization purpose, which can both guide the variables choice as well as shape how the variable changes over time.

Table 5.2 Summary and re-elaboration of Panëels and Roberts (2010) review

Category	Sub-category	Description	Case study
Data type	Scalar data	Represent single-valued data points using force feedback or texture variations	ListeningCups (Desjardins and Tihanyi 2019), Laina (Menheere et al. 2021), Equipoise
	Vector data	Use directional feedback to represent data with magnitude and direction	Sensbiom, Loop (Sauvé et al. 2020)
	Multivariate data	Combine multiple sensations for datasets with multiple attributes	Laina, Air Transformed (Quick & Posavec 2015)
Interaction mode	Exploratory Haptics	Allow users to 'explore' the dataset by moving and sensing variations in a feedback	DataStrings, DataChest (Wijers et al. 2024), ListeningCups, Air Transformed, Motiis (Pepping et al. 2020)
	Active interaction	Users intentionally manipulate the device to probe the data	DataStrings, DataChest
	Passive interaction	Users receive feedback while performing another task	Sensbiom, Laina, Bloats (Besana 2021)
Visualization purposes	Quantitative understanding	Deliver precise information through measurable variations	Equipoise, Pulse (Kison 2008), Drum Roll (Waldschütz et al. 2020)
	Qualitative insights	Convey general trends or relationships rather than exact values	Sensbiom, Loop
	Relational exploration	Help users understand relationships between data points or sets	DataStrings, DataChest, Loop

Following the annotation phase, a first analysis we performed on the portfolio was mapping the case studies over two progression lines, a first one defining the data encoding level into the artefact (present/absent), a second one on the interaction with users (present/absent). We placed each case into the lines and this step provided very first intuitions on the data encoding level and the user interaction. Nonetheless we shortly realized that such kind of representation did not allow to spot the correlation between data encoding and user interaction.

Thus, we needed more established concepts to perform the analysis. From that moment on, we started referring to the notions of feedback and feedforward when identifying the interaction schema of a physicalization.

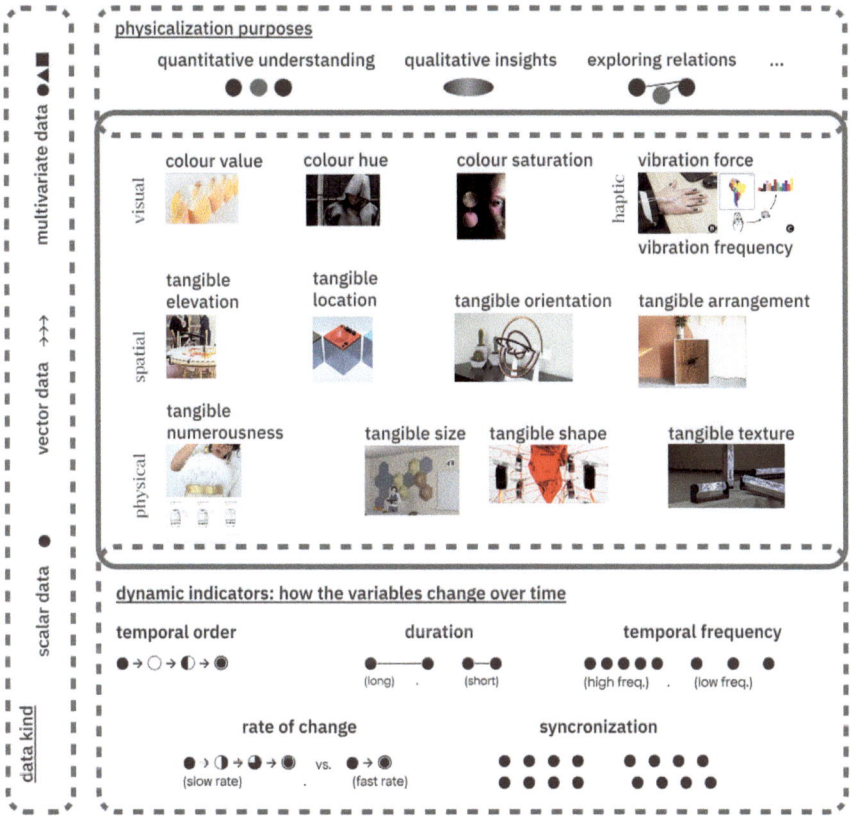

Fig. 5.1 Variables scheme

5.3 Feedback and Feedforward: An Analysis of the Annotated Portfolio

As mentioned, the annotated portfolio provided several insights about forms and aims of data physicalization. We propose a way to extract insights about the user interaction by looking at feedback and feedforward of data phys cases.

The concepts of feedback and perceived affordance have become foundational in the field of interaction design, particularly following Donald Norman's influential work "The Design of Everyday Things". Expanding on these ideas, Djajadiningrat et al. (2002) introduced the notion of feedforward to the human–computer interaction (HCI) community. They defined feedforward as the communication of an action's purpose, suggesting that this concept may be more suitable for electronic and interactive systems than affordance alone. In their view, a system is well-designed when it provides clear feedforward and inherent feedback, ensuring a proper coupling between user actions and system responses. Later, Vermeulen et al. (2013) further explored feedforward as a mechanism to address the cognitive gap between a user's intention and the system's response. Briefly, while perceived affordances signal the availability and method of interaction (i.e., how to perform an action), feedforward communicates the expected outcome of that action, and what will happen when the user proceeds.

We decided to perform an additional activity on the annotated portfolio by framing the case studies through the lens of feedback and feedforward. We submitted our initial findings to MSc students in Digital and Interaction Design, of the School of Design of Politecnico di Milano, in December 2023. Together with the students we placed the case studies into a matrix having on the X axes the feedback (from absent to present) and on the Y axes the feedforward (from perceptible to hidden). Then, we discussed the matrix together to find correlations, contradictions, and draw insights.

5.3.1 Feedback

During the discussions with students, one recurring theme was the ambiguous nature of feedback in static data physicalizations like Air Transformed and ListeningCups. These artefacts do not visibly change in response to data, leading some to view them as lacking feedback altogether. Others, however, suggested a broader understanding—where feedback is not absent but rather extended across time, unfolding between the creation and eventual disuse of the artefact.

Dynamic data physicalizations, in contrast, were seen as offering a more immediate connection between input and output, yet still not always easy to interpret. Projects such as Laina, which slowly translates running data into pin movements on a board, were perceived as unclear in conveying value or meaning, raising the question: "Did I perform well or not?". Nonetheless, such objects encourage connection and interdependence, leading to forms of affection, where the data are not represented in a symbolic equivalent of graphs and charts, but in an ephemeral, undirect way.

Among dynamic examples, those relying on simple data and visual changes such as colour were seen as the most comprehensible. The Sensbiom installation, which transitions from orange to red in response to UV radiation levels, was frequently cited for its effective real-time feedback.

Constructive data physicalizations emerged as particularly strong in terms of feedback clarity. Here, the user's action directly builds the representation. For example, in Data Strings by Domestic Data Streamers, the public manipulates physical strings to visualize data, with cause and effect occurring nearly simultaneously. This aspect bridges the gap between action and feedback in a tangible, immediate way.

5.3.2 Feedforward

The feedforward, intended as what will happen when a user performs an action on the object, seems perceptible for the static physicalizations. In fact, they are everyday artefacts (a necklace and cups), so the user expects to use them as any other resembling object.

Similarly, it seems perceptible for constructive data phys. For instance, the DataChest case is a constructive toolkit for self-tracking, enabling users to visualize and make sense of their data using tokens of different sizes and shapes (Wijers et al. 2024). Students agreed on the fact that the result of the action might be exactly what the user expects, since they know beforehand what kind of physicalization they intend to construct. Despite this insight may be correct, a recent study from Daneshzand and colleagues (2025) highlights how the design choices taken while constructing the physicalization might generate unpredictable and unexpected physicalization results.

Dynamic ones instead tend to have hidden feedforward. This aspect unfolds in two stances. In the first, user does not have the possibility to interact with the artefact. Examples in this sense are Sensbiom (UVR exposure, see above) or See Boat (MIT Media Lab, 2020), which are remote controlled boats that measure and display water quality on site and in real-time. In both cases, the user is an observer, and they might respond to the communicated data in their everyday life, through behavioral change, for instance protecting themselves more carefully from UVR or sensibilize their community to the issue of water pollution.

In the second stance, user might interact with the artefact, but it is not clear how and what they can expect to happen. For instance, Motiis is a system that tangibly visualizes children's emotions experienced during gaming sessions (Pepping et al. 2020). Parents can haptically interact with it to feel the data, nonetheless the effects of their actions are not clear: will the artefact suggest ways for the parents to open conversations with their children? Will it record the different emotions?

Students agree that the dynamic physicalizations with the most perceptible feedforward is the case of Bloats, a flexible family of smart wall modules that monitor and communicate the indoor air quality (IAQ) of school classrooms (Besana 2021). Each module has a specific call to action which clarifies to teachers and students how to behave, while educating them to pay better attention to IAQ issues.

5.4 Applying the Variables Scheme in Speculative Artefacts

In a subsequent phase, we applied the variables and metaphors scheme in a hands-on activity conducted with MSc students in Digital and Interaction Design, in November 2024. The goal was to assess whether the materials previously formalized through the literature review and annotated portfolio could serve as foundational tools for ideating data physicalizations, ideas that that could later be prototyped and used as research probes in real-world user enagagements.

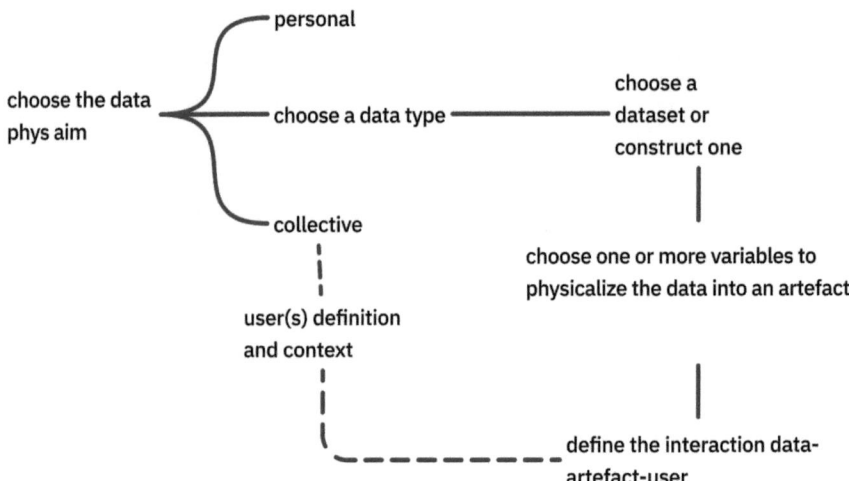

Fig. 5.2 Preliminary process to ideate physicalizations, tested with students

We introduced the scheme as a starting point for designing artefacts that embody autographic data, with a focus on the context of menstrual health. The session began with a discussion of how data can be embodied through encoding variables and metaphors, followed by an exploration of feedback and feedforward concepts. We then moved into the practical phase of the activity.

To support the ideation process, we presented both a process to ideate physicalizations and the variables + metaphors scheme (Fig. 5.2). Starting from the broader topic, students identified a specific user group. They then selected relevant data types, chose appropriate encoding variables to translate those data, and finally explored how to incorporate the physicalization into either a mundane, familiar object or a newly conceived artefact. One constraint we gave to the students is that their physicalization idea should be either constructive or dynamic, in any case enable an active user interaction over time.

The activity was conducted using a shared Figma board. Students were given six case studies from the annotated portfolio, each related to menstrual or reproductive health data. Additionally, the board included a structured overview of menstrual cycle data types, reflecting the common metrics used in popular menstrual tracking apps: menstrual phase, flow, flow intensity, flow colour, spotting, and symptoms (rated on a 0–5 scale) such as cramps, fatigue, breast pain, headache, basal body temperature, and cervical mucus quality.

To support ideation further, we also provided four pre-prepared datasets combining different data types, which students could use as a basis for their physicalization ideas.

5.4 Applying the Variables Scheme in Speculative Artefacts 41

5.4.1 Results

The students worked in groups and shared their concepts using a canvas we had prepared on the shared Figma board. The canvas was designed to help them articulate their ideas following the process provided them, by prompting them to specify whether their physicalization was intended for personal or collective use, to identify the user, define the data or data types used, indicate the encoding variables, and describe the concept—optionally using evocative images. Lastly, we asked them to reflect on the potential impact of their idea on the selected user. Among the ones generated during the ideation activity, three ideas illustrate different approaches to menstrual and reproductive health through data physicalization.

1. Genesis is a collective artefact designed to indicate optimal fertility periods for conception by tracking cervical mucus quality. The system consists of a reusable stick for analyzing cervical mucus, which transmits data to a pair of stones, one for each partner. These stones light up in different colours depending on the likelihood of pregnancy, with hue, saturation, and brightness encoding the fertility level. Vibration force is also used to convey feedback. The stones can be shaken to recall the most recent result, encouraging mutual awareness and communication between partners. Data are stored in a shared database, and the artefact promotes involvement from both individuals.
2. CuddleCloud is a personal artefact that supports menstrual well-being through stress-relieving interaction. The object is soft, warm, and malleable, designed to be squeezed, stretched, or folded, providing comfort and tactile engagement during menstruation. The artefact encodes data related to menstrual phase, pelvic pain, and flow colour using tangible shape and colour saturation. By interacting with the object, users find relief from discomfort while subtly inscribing their experience into the materiality of the artefact, turning personal pain into a traceable, embodied dataset.
3. Tide is a collective artefact aimed at raising awareness of vaginal discharge among individuals in early menopause. The concept revolves around a wearable ring that rotates slowly over the course of a month. The ring's surface integrates magnetic powder that changes its colour, texture, and shape in response to changes in cervical mucus quality and vaginal dryness. Tangible orientation correlates with the menstrual calendar, providing a layered sensory feedback system. The artefact translates intimate bodily data into a slow, continuous material transformation, offering a discreet and reflective tool for users navigating changes in their reproductive health.

5.4.2 Key Findings

The process proved helpful in supporting students in the early ideation of physicalizations, enabling them to define the data type, dataset, and user(s) with some clarity. Interestingly, most of the concepts proposed were collective physicalizations, shared with a partner, or, in one case, with coworkers, thus revealing a social dimension. At this early stage providing ready-made datasets may be premature. In fact, most students worked with general data types rather than the datasets themselves, suggesting that datasets may become more relevant later, during the prototyping phase.

Notably, students were able to combine data types in novel ways. For example, one group proposed a physicalization that merged menstrual flow colour and cervical mucus quality, a data pairing not originally envisaged. The canvas itself was a useful structuring tool, although we noted that the idea box could have been further broken down to guide the communication more effectively. Many students naturally situated their concepts within specific use contexts, an observation that resonates with current literature, as well as the physicalization case studies, underscoring the importance of context and situatedness in data physicalization.

Among the variables used, changes in colour and tangible shape were the most common, perhaps due to their familiarity or alignment with embodied metaphors (e.g., a light-to-dark colour gradient representing increasing intensity or severity, such as pain). Variables such as tangible orientation or elevation were less frequently used, and tangible numerousness was not used at all.

After presenting their ideas, students engaged in a collective discussion, during which we discussed each idea according to the feedback/feedforward lens (Fig. 5.3).

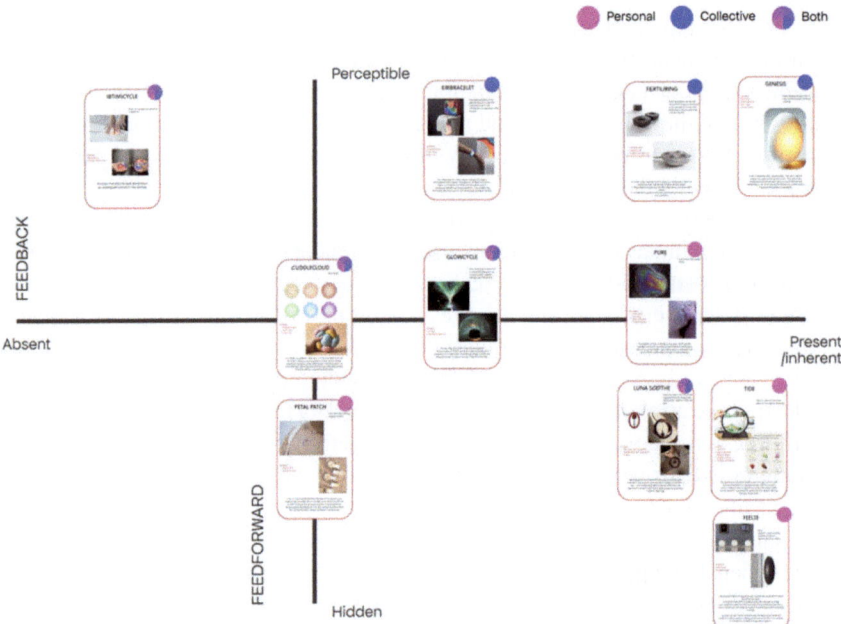

Fig. 5.3 Overview of students' ideas into the feedback/feedforward matrix

Considering that students were already familiar with the matrix, and the feedback and feedforward concept, they approached these concepts to frame the interaction. Most ideas have present feedback, some perceptible feedfoward. There is a correlation between personal and collective artefact, with the feedforward, either perceptible or hidden. In fact, collective physicalizations have perceptible feedforward, while personal physicalizations tend to have hidden feedforward. This might be because in collective feedforward the main aim is to connect partners in a shared activity regarding reproduction (such as ovulation tracking, or temperature sensing), while for personal physicalization it may be a device connected to a wearable, or an app, which showcases the data to the primary user, or person who owns those data; therefore does not interact directly with the object.

5.5 Conclusion

This chapter has explored the practical manifestations of data physicalization, presenting an annotated portfolio of diverse case studies and analyzing their interaction through the way data are encoded in the artefact and through the lens of feedback and feedforward. By examining real-world examples, we highlighted how data phys vary in aesthetic, aim, technique, and user interaction.

Mapping the annotated portfolio against design dimensions revealed the complexity of encoding variables when it comes to data. By integrating established frameworks, such as those from Panëels and Roberts (2010), with more recent insights from Ranasinghe and Degbelo (2023), we emphasized the need for customized approaches in data physicalization design.

Static, constructive, and dynamic data phys each offer unique affordances and challenges. Static ones prioritize aesthetic value and contemplation but lack real-time interaction and update over time. Constructive physicalizations encourage hands-on engagement and are ideal for collaborative decision-making, while presenting challenges due to the metaphors required to understand data. Dynamic physicalizations leverage cutting-edge materials and technologies to represent evolving datasets, though they often face challenges in ensuring clear feedback and feedforward.

Moreover, we present the results of a hands-on activity conducted with students, which involved using the variable and metaphor schema along with a process for ideating physicalizations. This activity offered valuable insights into the most preferred variables for data physicalization and highlighted how the different elements involved in the process interact and influence each other: the purpose of the physicalization, the user, and the data type. Which comes first, the aim, the user, or the data?

The schema and process proposed in the activity can effectively guide designers and researchers to ideate a first array of physicalizations for a specific context, to be then prototyped and used as design probes in focus groups, interviews, or co-design sessions. This very first step is useful to iterate on data types and variables, to explore potential physicalization aestethics (i.e., a data object or a fully new artefact), and interaction modalities. These tools also push designers and phys creators to think about the temporality of the physicalization. Reflecting on the concepts of feedback and feedforward within a user context allows for a clearer understanding of whether the data is represented as a reactive element (feedback) or as an anticipatory one (feedforward).

Ultimately, we highlight the potentialities of integrating and adapting already conceptualized framework to ideate and evaluate data physicalization designs, proving the efficacy of feedback and feedforward when dealing with matters such as data, and envisioning perhaps novel and unique concepts to approach this emerging area.

References

Desjardins A, Tihanyi T (2019) ListeningCups: a case of data tactility and data stories. In: Proceedings of the 2019 on designing interactive systems conference, pp 147–160

Djajadiningrat T, Overbeeke K, Wensveen S (2002) But how, Donald, tell us how? On the creation of meaning in interaction design through feedforward and inherent feedback. In: Proceedings of the 4th conference on designing interactive systems: processes, practices, methods, and techniques, pp 285–291

Dragicevic P, Jansen Y, Vande Moere A (2020) Data physicalization. Handbook of human computer interaction, pp 1–51

Gaver B, Bowers J (2012) Annotated Portfolios. Interactions 19(4):40–49

Hogan T, Hornecker E (2013, February). In touch with space: embodying live data for tangible interaction. In Proceedings of the 7th international conference on tangible, embedded and embodied interaction, pp. 275–278

Jansen Y, Dragicevic P, Isenberg P, Alexander J, Karnik A, Kildal J, Hornbæk K (2015) Opportunities and challenges for data physicalization. In: Proceedings of the 33rd annual ACM conference on human factors in computing systems, pp 3227–3236

Kaper HG, Wiebel E, Tipei S (1999) Data sonification and sound visualization. Comput Sci Eng 1(4):48–58

Norman D (2013) The design of everyday things: revised and expanded edition. Basic books

Paneels S, Roberts JC (2009) Review of designs for haptic data visualization. IEEE Trans Haptics 3(2):119–137

Pepping J, Scholte S, Van Wijland M, de Meij M, Wallner G, Bernhaupt R (2020) Motiis: fostering parents' awareness of their adolescents emotional experiences during gaming. In: Proceedings of the 11th Nordic conference on human-computer interaction: shaping experiences, shaping society, pp 1–11

Pennerup Nilsson A (2023) Unintentional artefacts: recycling data through the looking-glass

Ranasinghe C, Degbelo A (2023) Encoding variables, evaluation criteria, and evaluation methods for data physicalisations: a review. Multimodal Technol Interact 7(7):73

Signer B, Ebrahimi P, Curtin TJ, Abdullah AK (2018) Towards a framework for dynamic data physicalisation. In: Proceedings of the international workshop toward a design language for data physicalization, Berlin, Germany

Sosa R, Gerrard V, Esparza A, Torres R, Napper R (2018) Data objects: design principles for data physicalisation. In International Design Conference 2018, pp. 1685–1696. University of Zagreb

Vermeulen J, Luyten K, van den Hoven E, Coninx K (2013) Crossing the bridge over Norman's Gulf of execution: revealing feedforward's true identity. In: Proceedings of the SIGCHI conference on human factors in computing systems, pp 1931–1940

Wang Y, Ma X, Luo Q, Qu H (2016) Data edibilization: representing data with food. In: Proceedings of the 2016 CHI conference extended abstracts on human factors in computing systems, pp 409–422

Wijers J, Brombacher H, Houben S (2024) DataChest: a constructive data physicalization Toolkit. In: Proceedings of the eighteenth international conference on tangible, embedded, and embodied interaction, pp 1–7

Zhao J, Vande Moere A (2008) Embodiment in data sculpture: a model of the physical visualization of information. In: Proceedings of the 3rd international conference on digital interactive media in entertainment and arts, pp 343–350

References of the case studies

Quick M, Posavec S, (2015) Air transformed. https://www.stefanieposavec.com/airtransformed
Besana N (2021) Bloats. https://www.politesi.polimi.it/handle/10589/186781
Wijers et al (2024) DataChest. https://doi.org/10.1145/3623509.3635252
Domestic Data Streamers (2014) Data strings. https://www.domesticstreamers.com/work/data-strings/
Waldschütz H et al. (2020) Drum Roll. https://doi.org/10.1145/3393914.3395848
Menheere D et al. (2021) Laina. https://doi.org/10.1145/3461778.3462041
Desjardins A, Tihanyi T (2019) ListeningCups. https://doi.org/10.1145/3322276.3323694
Sauvé K et al. (2020) Loop Lamp. https://www.kimsauve.nl/files/LOOP_NordiCHI2020.pdf
Wu S (2023) Menstrual calendar. https://shirleywu.studio/thoughts/2023/11/menstrual/
Pepping J et al. (2020) Motiis. https://doi.org/10.1145/3419249.3420173
Kison M (2008) Pulse. https://www.markuskison.de/pulse.html
MIT Media Lab (2020). https://www.media.mit.edu/projects/thermal-fishing-bob-in-place-environmental-data-visualization/overview/

Chapter 6
Looking Ahead: Which Emerging Directions?

Abstract This chapter outlines a preliminary framework for designing data physicalizations, combining conceptual insights with practice-based tools such as a variables scheme and ideation process. The chapter identifies three emerging directions (living data mapping, lean data engagement, and data ownership) that suggest new pathways for designing with data.

6.1 A First Framework Attempt

This book provides a theoretical and conceptual overview of data as material to design, outlining definitions, current practices and methodologies, and emerging areas worth of further exploration. Starting with a critical rethinking of what data is and how it is situated, intimate, and shaped by context, we moved from theoretical reflections to practical applications. We mapped key developments in tangible interaction design, including interfaces and materials that bridge the digital and physical. Through an extensive literature review, we identified foundational principles and the role of metaphors in data representation. We then grounded these insights in practice, using case studies and the method annotated portfolio to examine how interaction, feedback, and feedforward emerge in real-world data physicalizations. Together, these chapters offer a foundation for reimagining how data can be made not only visible, but experiential and embodied.

These activities informed the elaboration of two key tools to ideate physicalizations: a variables scheme and an ideation process. These two preliminary outputs will lay the foundation of a cohesive design framework for designing physicalizations, including a process encompassing ideation, prototyping and evaluating, as well as the tools to perform such activities. The development of such framework will follow a research-through-design approach of testing and iterating the tools. This diagram is a first adaptation of the Double Diamond process (Design Council 2019), reconfigured to support the specific challenges and opportunities of designing data physicalizations. It builds on the two preliminary outputs introduced in this book and aims to offer a structured yet flexible framework for ideation, prototyping, and implementation.

6.1.1 Variables Scheme

The process of data mapping is complex and multifaceted. This is particularly true for dynamic and constructive physicalizations as they evolve over time, according to both datasets and user's response. For this reason, the proposed variables scheme includes the temporality component, which should be carefully addressed as a design consideration, for instance through prototypes evaluation, user testing, and generative sessions (Fig. 6.1).

The scheme includes familiar variables and less familiar ones. Colour changes, whether it's hue, saturation, or value, draw on a long-lasting tradition within graphic and communication design, while haptics, physical, and kinaesthetic variables are less embodied into such practices. Their use in data physicalization might expand design and interaction possibilities, and even rethink how devices can communicate digital information, which is widely based on visual and sound modalities.

The scheme suggests different entry points for choosing a variable: the data kind, the visualization purpose, or the variable itself, perhaps as an applied exploration of a particular technology such as shape-changing interfaces. In a different scenario, if we are dealing with scalar data (single valued data points) and the physicalization purpose is quantitative understanding, using tangible arrangement or numerousness as variable might be ideal, since they are prompted by single units of materialization. This scheme is a starting point that will be iterated through a series of mini-workshops and experimentations.

6.1.2 Ideation Process

Data physicalization clearly represents a peculiar category of interactive artefact. Unlike tracking devices, it differs both in how data is embedded within the artefact and in the types of interactions it claims to enable, that is multimodal and multisensory. This raises a fundamental question, reminiscent of the classic chicken-and-egg dilemma: where should the design process begin? With the data itself, or with the intended purpose of the physicalization?

Framing the interaction with data physicalizations may require a direct engagement with users, particularly given the novelty of this artefact typology. In this sense, users should be actively involved in the physicalization designs. The process proposed here is intended as a foundational step to support the ideation of a wide range of physicalizations, which can then be crafted, prototyped, and used as design probes to explore user engagement (Fig. 6.1).

6.1 A First Framework Attempt

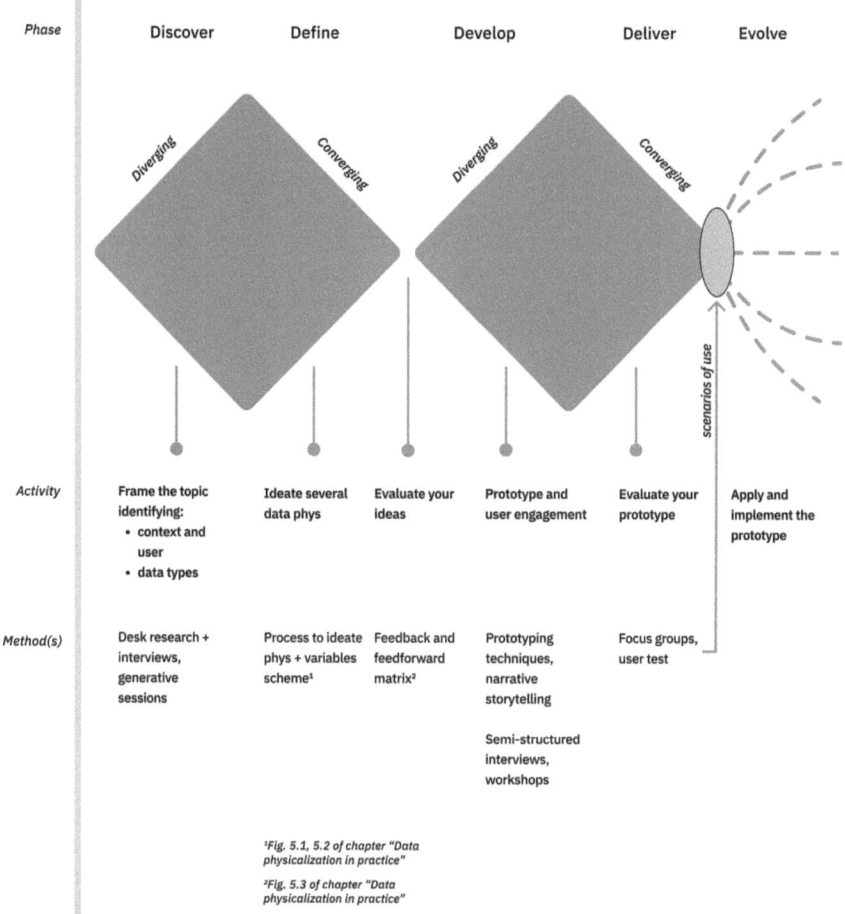

Fig. 6.1 Preliminary framework, adapted from the double-diamond process (Design Council 2019)

This phase can be seen as the first stage in a broader and more structured design process for data physicalizations, one that starts with ideation, moving to prototyping and then into evaluation. Much like traditional design models, including the double diamond framework, th initial phase is exploratory and generative. The tools presented aim to support the creation of early concepts that can subsequently be refined and developed for testing.

The next stage of this research will focus on iterating the proposed tools through a series of case studies, each addressing specific users and application domains. In parallel, further iterations on their efficacy will be carried out through a number of focused workshops and hands-on activities.

6.2 Emerging Directions

Three key directions emerge from this work, pointing toward future efforts at the intersection of materials science, STS, and human–computer interaction. First, the increasing use of materials with agency, such as bio-based substrates or AI-driven technologies, opens possibilities for more dynamic, living forms of data representation. These enable novel encoding and decoding patterns that evolve with time and context. A second direction concerns a novel meaning of interaction between data and bodies, where physicalizations are not just expressive outputs, but tools that encourage a more embodied and intuitive engagement with data. Such an approach supports a leaner mode of data interaction, less focused on reporting numbers, and more on guiding conscious behaviors and interpretations. Third, data physicalization has the potential to become a tool for more critical data practices. As tangible representations, data physicalizations can empower users to reclaim agency over their own data, shifting from passive consumption to active engagement. These three directions are explored below, each culminating in a question that we envision to be triggering and inspiring for practitioners and researchers in the field.

6.2.1 Living Data Mapping

The field of information visualization (infovis) and related fields have formalized faithful data encoding–decoding schemas and processes, using visual variables to represent data in the most accurate way. In infovis, data is represented through a combination of visual elements such as dots, lines, and areas, along with attributes like size, colour, and position. In its infancy, such processes have been transposed to the physicalization field as well, with the same intention of faithfully convey data. The literature review shows that several endeavours have been made to formalize *ad-hoc* vocabularies for data physicalization (Bae et al. 2022; Hornecker et al. 2023). These efforts come across as key starting points to give an identity to the field, distinguishing data phys from its closely related areas.

Inspired by philosophical approaches such as new materialism, data feminism, and agential realism, recent research strands have begun to investigate situated and indexical forms of data representation. Data becomes something diffractive, inextricably connected to the phenomena it aims to represent, and self-inscribing onto materials. Offenhuber (2020) refers to autobiographical visualization to indicate data that manifest itself leaving material traces. Design efforts might be directed to set the ideal conditions for data to manifest, using for instance smart or responsive materials, or instead practices of co-creation.

Therefore, how do we deal with diffractive, agential, and living data mapping, that shifts the focus from data accuracy to unpredictability and perhaps even *messiness*?

6.2.2 Lean Approach to Data

We generate vast quantities of data on a daily basis, flowing continuously through devices, networks, and cloud infrastructures. Much of this data ends up unused, and even the slimmer portion that is eventually communicated to users often fails to achieve its intended purpose. One reason is that data is frequently conveyed in overwhelming volumes, presented in abstract numerical form, or communicated in ways that feel detached and decontextualized.

In the field of interaction design, particularly in relation to tracking technologies, an alternative approach is emerging that embraces a more selective, thoughtful engagement with data. This lean approach prioritizes clarity over quantity by reducing the amount of data presented at once, and by leveraging combined datasets interpreted through machine learning or artificial intelligence to convey more meaningful insights.

A compelling example of this approach is OTO by the design studio Takram. OTO is a tracking device and scale for cycling enthusiasts that avoids presenting charts or raw performance metrics. Instead, it distills the data into short, meaningful prompts such as *"you are well"* or *"remember to stay hydrated,"* which support real-time reflection and action.

Similar tendencies can be observed in data visualization and information design, where minimalism is increasingly adopted to enhance usability. This shift is also influencing the fields of user interface (UI) and user experience design (UXD), particularly in the design of apps and digital dashboards in general.

Despite these developments, the lean approach remains underexplored in the context of data physicalization. This raises important questions: how might we move beyond the idea of simply *"making data physical"* to instead create physicalizations that foster reflection, suggest action, and promote empowerment?

6.2.3 Physicalizations and Data Ownership

Critical perspectives on digital apps and, more broadly, digital products that engage with personal data often centre around privacy and ownership. This is largely due to the fact that data is typically sensed by wearable technologies or other connected devices and then transmitted to digital platforms, ultimately stored in cloud infrastructures. This model of data tracking runs the risk of individualizing and privatizing aspects of health, especially when free-to-use apps request access to users' personal data in exchange for tracking services. While more sophisticated technologies may offer improved data security, they often remain less accessible, and exclusive.

Data physicalization has the opportunity to offer a compelling alternative by promoting more tangible and embodied relationships with personal data. Through physical forms of monitoring, measuring, and autobiographical tracking of personal data (whether health-related or otherwise), data can remain within the control of the individual. In such contexts, people decide how and with whom to share their data, and physicalizations can act as mediators in this exchange, supporting more autonomous data practices.

This shift opens space for emerging practices such as data donation, where data are no longer seen as sources of disembodiment or vulnerability but rather as something meaningful, even playful, to explore. In the context of data physicalization and design research, data donation refers to the voluntary act of sharing one's personal data (often autobiographical or self-tracked) for the purpose of design exploration, prototyping, or participatory inquiry. It is not about data being extracted, but rather about the intentional, situated, and often collaborative use of data, especially in research or co-design settings. These approaches encourage users to engage with their data in ways that foster curiosity, reflection, and agency, moving away from the alienating tendencies often associated with datafication.

6.3 Final Remarks

This closing chapter has outlined the main contributions of the book by drawing together insights from both literature and practical case studies. Particular attention has been given to the two preliminary research outputs: the variables scheme and the ideation process for physicalization. These tools are intended to be open, to be refined through iterative, research-through-design practices and further tested in workshops and other applied contexts.

In parallel, the chapter has highlighted a set of emerging research directions that have surfaced through the interdisciplinary intersections of design, science and technology studies (STS), and human-data interaction. These include exploring living data mapping, adopting a leaner approach to data representation, and reflecting on issues of data ownership, potentially enhanced by physicalization practices.

Rather than offering definitive answers, the book aims to provide a conceptual and practical foundation for those interested in engaging with data as a material to and for design. We hope it can serve as a reference for further experimentation and reflection, both by offering tools and processes, and by prompting new questions around how data can be experienced, shared, and situated in meaningful ways.

References

Bae SS, Zheng C, West ME, Do EYL, Huron S, Szafir DA (2022) Making data tangible: a cross-disciplinary design space for data physicalization. In: Proceedings of the 2022 CHI conference on human factors in computing systems, pp 1–18

Design Council (2019) Beyond the double diamond: a universally accepted depiction of the design process. https://www.designcouncil.org.uk/our-resources/the-double-diamond

Hornecker E, Hogan T, Hinrichs U, Van Koningsbruggen R (2023) A design vocabulary for data physicalization. ACM Trans Comput-Hum Interact 31(1):1–62

References

Offenhuber D (2020) What we talk about when we talk about data physicality. IEEE Comput Graph Appl 40(6):25–37

OTO (2018) Better decisions with lean data, Studio Takram. https://www.takram.com/ja/podcasts/better-decisions-with-lean-data

MIX
Papier aus verantwortungsvollen Quellen
Paper from responsible sources
FSC® C105338

If you have any concerns about our products,
you can contact us on
ProductSafety@springernature.com

In case Publisher is established outside the EU,
the EU authorized representative is:
**Springer Nature Customer Service Center GmbH
Europaplatz 3, 69115 Heidelberg, Germany**

Printed by Libri Plureos GmbH
in Hamburg, Germany